LIFE, WAR, EARTH

LIFE, WAR, EARTH

Deleuze and the Sciences

John Protevi

University of Minnesota Press

MINNEAPOLIS · LONDON

See publication history on page 247.

Published by the University of Minnesota Press
111 Third Avenue South, Suite 290
Minneapolis, MN 55401-2520
http://www.upress.umn.edu

Library of Congress Cataloging-in-Publication Data

Protevi, John.
  Life, war, earth : Deleuze and the sciences / John Protevi.
  Includes bibliographical references and index.
  ISBN 978-0-8166-8101-3 (alk. paper)
  ISBN 978-0-8166-8102-0 (pbk. : alk. paper)
1. Deleuze, Gilles, 1925–1995. 2. Science—Philosophy. I. Title.
  B2430.D454P76 2013
  194—dc23
                        2013006824

Printed in the United States of America on acid-free paper

The University of Minnesota is an equal-opportunity educator and employer.

20 19 18 17 16 15 14 13   10 9 8 7 6 5 4 3 2 1

Contents

Preface

This book is based on work produced from 2008 to 2012, that is, during and after work on *Political Affect* (2009). That book in turn built on the theoretical background established in *Political Physics* (2001) and *Deleuze and Geophilosophy* (with Mark Bonta; 2004). I will recapitulate some of that background in the first section of the introduction to this book, but readers interested in more full explications of the ideas should consult the earlier works.

Many of the chapters presented here were written as independent pieces for journals or essay collections; others were delivered as talks but not published. I have reworked all the pieces to reduce repetition, to refine formulations, to establish a consistent vocabulary, and to introduce cross-references. The most important of these moves is to present in the first section of the introduction a sketch of Deleuze's ontology, which is invoked in all the pieces. Despite the reworking, the level of the formality of presentation—and perhaps, though I hope not too much, the rigor of the argument—will vary in relation to the genesis of the chapter.

My approach is highly speculative, but I hope the concepts are empirically responsible. That is, in sketching the interlocking dimensions of the various multihyphenated multiplicities I invoke (the "geo-hydro-solar-bio-techno-political"), I pay close attention to the earth, life, and cognitive sciences. However, I do not try to intervene in specialist debates, so the references will sometimes be to syntheses and overviews rather than to specialist contributions. I hope that the book will serve to introduce scientifically minded philosophers and philosophically minded scientists to the benefits of a Deleuzian approach and to introduce Deleuzians to the possibilities of linking his thought to that of some of the most interesting and important of the contemporary sciences.

Acknowledgments

There are many people to thank for helping me formulate these ideas. No such list could be complete, but let me especially thank the following and ask for forgiveness for anyone omitted: Manola Antonioli, Jeff Bell, Miguel de Beistegui, Rosi Braidotti, Manuel Cabrera Jr., Andy Clark, Jon Cogburn, Amy Cohen, Claire Colebrook, William Connolly, Dock Currie, Dennis Des Chene, Hanne De Jaegher, Manuel DeLanda, Ezequiel Di Paolo, Chuck Dyke, Shaun Gallagher, Liz Grosz, Lisa Guenther, Eugene Holland, Joe Hughes, Mark Lance, Len Lawlor, Todd May, Jeff Nealon, Alva Noë, Catarina Dutilh Novaes, Susan Oyama, Paul Patton, Roger Pippin, Patricia Pisters, Anne Sauvagnargues, Eric Schliesser, Hasana Sharp, Jan Slaby, Nick Srnicek, Evan Thompson, Dan Smith, Alistair Welchman, Mike Wheeler, Cynthia Willett, Liz Wilson, and James Williams. Special thanks are due to Mark Ohm for research assistance, and thanks go as well to the entire New APPS Borg for wonderful intellectual companionship.

Deleuze and the Sciences

The subtitle of this book is "Deleuze and the Sciences," so I will say how I see that relation in this section of the introduction. First, we can note that the very idea that there is a positive relation runs contrary to a widely held belief identifying the mainstream of twentieth-century French thought with a suspicion of science coupled with a commitment to the "end of metaphysics" or "end of philosophy."[1] However, Deleuze in fact sees himself as providing a metaphysics of contemporary science. In a very clear self-description, Deleuze (as cited in Villani 1999, 130) says, "I feel myself to be a pure metaphysician. . . . Bergson says that modern science hasn't found its metaphysics, the metaphysics it would need. It is this metaphysics that interests me."

I am not alone, of course, in sympathetically discussing Deleuze and the sciences. Since the early 1990s, a number of works (among others, Massumi 1992; DeLanda 2002; Bonta and Protevi 2004; Beistegui 2004; Bell 2006) have claimed that Deleuze offers a naturalist ontology that maps well onto wide-ranging current research projects that use nonlinear dynamic systems modeling.[2] The scientific fields using these techniques are now widespread, from geomorphology and meteorology in the earth sciences to ecology and genomics in the life sciences, economics and sociology in the social sciences, and neurodynamics and developmental biomechanics in the cognitive sciences. The utility of this approach comes from the way Deleuze's ontology can help us think of individuation as the integration or resolution of a distributed and differential system, that is, a system in which multiple processes interact such that qualitative changes in the behavior of the system occur at singular points in the relation of their rates of change. For a relatively simple meteorological example, consider how, at a singular point, the relation of the rates of change of temperature, air pressure, air circulation, water vapor concentration, condensation, evaporation, and so on, will be such that raindrops form and fall. Let us now examine the relation of Deleuze and dynamic systems theory in more detail.

Deleuze and Dynamic Systems Theory

I will here present a simplified sketch, based on the presentation in *Political Affect* (Protevi 2009), of dynamic systems theory, and then show how we can read Deleuze in relation to it. There are three sets of terms to distinguish here: (1) in the system being modeled, range of behavior, fluctuation, patterns, and thresholds; (2) in the dynamic model, phase space, trajectory, attractors, and bifurcators; and (3) in the mathematics used to construct the model, manifold, function, and singularity. A *phase space* is an imaginary space with as many dimensions as the variables of a system chosen for the model. The phase space model is constructed using a manifold, an *n*-dimensional mathematical object. The manifold qua phase space represents the range of behavior open to the system. The condition of the system can be represented by a point in phase space with a value for each dimension. Tracked over time, the point traces a trajectory through the phase space, a trajectory representing the behavior of the system. Points toward which trajectories asymptotically converge are called *attractors* and represent patterns of behavior of the real system. The areas of phase space surrounding attractors—representing normal behavior of the system in one or another of its behavior patterns—are called *basins of attraction*. The behavior patterns described by attractors are, in highly complex (biological and social) systems, formed by the action of negative feedback mechanisms, which are to be contrasted to positive feedback loops, which, instead of returning a system to a homeostatic set point, set up runaway growth or decline, which often pushes the system to adopt another behavior pattern. Positive feedback loops are thus represented by *bifurcators,* which model the points at which systems jump from one pattern of behavior to another, that is, in model terms, move from one basin of attraction to another. Positive feedback loops can also in some cases nudge a system to produce new behavior patterns, which would be represented by a new layout of attractors.

The layout of attractors and bifurcators in the phase space, which describes the layout of the patterns of behavior of the system, is defined by the layout of singularities, which are mathematical objects that define the topological structure of the manifold; a singularity is a point at which the graph of the function changes direction as it reaches local minima or maxima or, more dramatically, at which the slope of the tangent to the graph of the function becomes zero or infinite. A singularity in the manifold indicates a bifurcator in the phase space model, which in turn represents a threshold where the real system changes qualitatively. A singularity is not an attractor, but the distribution of singularities defines where attractors are found by indicating the limits of basins of attraction.[3] When systems are poised at the edges of one of their behavior patterns (when they cannot quite "make up their minds"), we find a zone of sensitivity in which minimal fluctuations—those that would be otherwise be swallowed up by the negative feedback loops whose operation defines normal functioning of a system following one of its patterns of behavior—can push the system into a new pattern or perhaps even to develop new behavior patterns. In modeling zones of sensitivity or crisis situations, we find fractal borders between basins of attraction so that any move, no matter how small and no matter what direction, might—or might not—trigger the move to another basin of attraction or even the creation of a new attractor layout.[4]

As we have said, what keeps complex biosocial systems inside a behavior pattern—represented by the trajectories inhabiting a basin of attraction—is the operation of negative feedback loops that respond to system changes below a certain threshold of recuperation by quickly returning the system to its pattern. These changes can be either endogenous fluctuations or responses to external events. Quickly recuperating systems are called *stable*. With regard to normal functioning, fluctuations or external events are mere "perturbations" to be corrected for in a stable system. Now changes of a certain magnitude—beyond the recuperative power of the negative feedback loops or homeostatic mechanisms—will push the system past a threshold, perhaps to another pattern in its fixed repertoire or perhaps into a zone where there are no patterns. Thus some stable systems are "brittle": they can be broken and die. Some systems are "resilient," however: a trigger that provokes a response that overwhelms its stereotyped defensive patterns and pushes the system beyond the thresholds of its comfort zones will not result in death but in the creation of new attractors representing new behaviors. We can call this "learning," although, of course, there is a sense in which the old system has died and the new one is reborn. Sometimes this creation of new patterns for a particular system repeats patterns typical of systems of its kind; we call this *normative development*. Sometimes, however, this change of patterns is truly creation: we can call this *developmental plasticity* (chapter 10).

To see the way Deleuze helps us see some of the philosophical significance of these notions, I will present a simplified sketch of the ontology found in *Difference and Repetition* (Deleuze 1968; 1994).[5] In this attempt at a "philosophy of difference" (Deleuze 1994, 29), Deleuze seeks a poststructuralist reformulation of Kant: an account of the transcendental conditions by which individuated entities (a person, a hurricane, a perception) are "genetically" produced by the actualization of a virtual field, or the integration of a differential field, or the resolution of a problematic field—the three expressions are synonymous for Deleuze (1994, 211).[6] The philosophy of difference counters many forms of what we might call "identitarian" philosophy, from Plato and Aristotle to Kant and Hegel and others, in which identities are metaphysically primary and differences are seen within a horizon of identity. With regard to Kantian transcendental philosophy, Deleuze (1994, 170) attempts to replace the Kantian project of providing the universal and necessary conditions for any rational experience with an account of the genesis of "real experience," the "lived reality [*réalité vécue*] of a sub-representative domain" (Deleuze 1968, 95; 1994, 69). Since Deleuze criticizes Kant's "tracing" operation, whereby the latter grounded empirical identities in transcendental ones (e.g., the Transcendental Unity of Apperception provides the ground for the unity of empirical psychological life), he demands a purely differential transcendental field. Taking his clue from Bergson, Deleuze names the ontological register of such a purely differential transcendental field "virtual" (Deleuze 1991, 95; 1994, 207). In *Difference and Repetition*, then, we find a tripartite ontological scheme positing three interdependent registers: the virtual, the intensive, and the actual.[7] Deleuze's basic notion is thus a tripartite "ontological difference": in all realms of being, (1) intensive individuation processes follow—but also in turn change, via "countereffectuation"—the structures inherent in (2) differential virtual multiplicities to produce (3) localized and individuated actual substances with extensive properties and differenciated qualities.

Simply put, then, the actualization of the virtual, that is, the production of the actual things of the world, proceeds by way of intensive individuation processes. In a fuller picture of Deleuze's ontology, we see that the virtual field is composed of "Ideas" or "multiplicities," which are constituted by the progressive determination of differential elements, differential relations, and singularities; what are related are precisely intensive processes, thought of as linked rates of change (Deleuze 1994, 182–91).[8] Beneath the actual (any one state of a system), we find "impersonal individuations" or intensive morphogenetic processes that produce system states, and beneath these, we find "preindividual singularities" (i.e., the key elements in virtual fields, marking system thresholds that structure the intensive morphogenetic processes). We thus have to distinguish the intense impersonal field of individuation and its processes from the virtual preindividual field of differential relations and singularities that make up an Idea or multiplicity. In a further nuance, we are able to distinguish *individuation* as the field of individuation from *dramatization* as the process of individuation. Thus the "order of reasons" from virtual through intensive to actual goes like this: "differentiation–individuation–dramatization–differenciation (organic and specific)" (Deleuze 1994, 251; 1968, 323; translation modified to restore the placement of the parentheses in the original).[9]

But what links the individuation field and the dramatization process? What starts things rolling, what triggers dramatization? Deleuze speaks of a "dark precursor" that connects different factors of the field of individuation and hence triggers "internal resonance" and then "forced movement" (Deleuze 1994, 126, 277–78). The first term, *dark precursor,* indicates information transfer between heterogeneous series (DeLanda 2002, 80). Deleuze (2004, 97) writes in "The Method of Dramatization" (the rough draft of chapter 5 of *Difference and Repetition*) that

> since intensity is difference, differences in intensity must enter into communication. Something like a "difference operator" is required, to relate difference to difference. This role is filled by what is called an obscure precursor [*précurseur sombre*; Patton

translates this as "dark precursor" in Deleuze (1994)]. A lightning bolt flashes between different intensities, but it is preceded by an *obscure precursor,* invisible, imperceptible, which determines in advance the inverted path as in negative relief, because this path is first the agent of communication between series of differences.

Deleuze seems to be unnecessarily gnomic with his term *dark precursor,* but in fact, it is drawn from mundane descriptions of meteorological processes. Actually, there are a number of steps here in lightning genesis: (1) the formation of ionized air, called "plasma," which is much more highly conductive than normal air; (2) the formation of step leaders, or channels of ionized air that propagate from cloud to ground in stages; (3) the formation of positive streamers, which reach from objects on the ground to clouds; and (4) the meeting of positive streamers and step leaders, which allows the current to pass. So *dark precursor* could be either step 2 or step 3. In fact, the term *précurseur* is used straightforwardly in any number of French-language websites on lightning.

Thus a simple example distinguishing and then linking the field and process of individuation can be found in the meteorological register, where the *field* of individuation is composed of the cloud–ground system, with its electrical potential differences, whereas lightning is the *process* of individuation, the production of an event, triggered by the dark precursor of step leaders and positive streamers. On a slower temporal scale, the field of individuation of a weather system would be the bands of different temperature and pressure in air and water that exist prior to and allow for the formation of wind currents or storms, which are the spatiotemporal dynamisms, the *process* of individuation of a singular event, sometimes worthy of its own name, as with hurricanes. In the biological register, which we will examine in chapter 10, an example of the field of individuation is the egg, while the process of individuation is embryogenesis, triggered by fertilization, which plays the role of the dark precursor by putting the different orders of genetic recombination, cell division, cytoplasmic gradients, and ultimately, gene expression in communication. However, to save Deleuze from tracing empirical individuation back to a transcendental identity qua "genetic program," we must see the biological virtual as the differential

Idea of genetic *and* epigenetic factors, as does the contemporary school of thought known as developmental systems theory.[10] The term *Idea* should not be seen in a Platonic sense. So let us pause to clarify that the Deleuzian virtual is non-Platonic, in two senses. First, it is not wholly separated from the actual; rather, there is two-way traffic between virtual and intensive such that the interaction of intensive processes changes the virtual conditions for future processes.[11] An example here would be a Deleuzian understanding of niche construction: the activities of organisms change the selection pressures for future generations. The ecological web of relations that we describe as "selection pressures" is not ghostly, it is perfectly real, but for Deleuze, it does not have the same ontological status as a single individuated act (e.g., a predator devouring a prey animal). Rather, the web is virtual, that is, composed of the relations of dynamically interactive processes. The virtual field is not composed of the processes themselves but by the differential elements, relations, and singularities of the processes. These elements, relations, and singularities are progressively determined by intensive individuation processes so that at critical points in the relation of predator and prey activity—at a singular point in the linkage of the rates of change of those processes—we can find the triggering of an event such as a population explosion or, in the opposite direction, an extinction.[12]

Second, the Deleuzian virtual realm is not composed of self-identical essences. Ideas are not sets of necessary and sufficient conditions for membership in a group, and they are not then instantiated by particulars. Rather, Ideas are "differentiated" as zones of intensity in a "space" of continuous variation. As such, they are "perplicated" or interwoven, and they blend into each other at their edges in what Deleuze (1994, 187; see DeLanda 2002, 22) calls "zones of shadow." A simple example of the distinction between essential difference and virtual differentiation would be tropical cyclones: the nominal definition distinguishing a hurricane from a tropical storm is sustained wind speeds of more than seventy-four miles per hour. But this is not reflective of any real distinction; it does not map onto physical turning points in storm formation but has only to do with the properties of already formed storms (this classification by properties is termed *differenciation*). For Deleuze, there is instead the Idea of tropical cyclones, but this is a continuous variation in which the "sub-Idea" of "thunderstorm cells" blends into that of "tropical depression," and that in turn blends into "tropical storm" and "hurricane." Continuous variation and perplication of Ideas does not deny that there are singularities or turning points in intensive processes leading to actual products. There are indeed singularities involved in the morphogenesis of tropical cyclones, but they have to do with temperature and pressure differences among wind and water currents triggering updrafts, eyewall formation, and so on; none of these relational points are captured by the classificatory notion of sustained wind speed. The latter is a matter of classification by property of an already formed system or differenciation. Deleuze, by contrast, is interested in the structure of the morphogenetic process leading to tropical cyclones. The question "which singular points formed this eyewall?" is a morphogenetic question posed in the virtual register of differentiation, whereas "what wind speed distinguishes a hurricane from a tropical storm?" is a merely classificatory question posed in the actual register of differenciation.[13]

A more complex example of perplication comes from biology. Deleuze indicates three dimensions of continuous variation among Ideas; along one of these dimensions, he will offer the example of "the varieties of animal ordered from the point of view of unity of composition" (Deleuze 1994, 187). That is, all animals share a unity of composition in the sense of being all composed of elements of nucleic acids and amino acids that enter into differential relations determining gene expression and protein synthesis. There is thus an Idea of "animal" that is expressed in different actualizations of those relations and their singular points. (Technically speaking, Deleuze will say that the relations are incarnated as qualities or "species" and the singular points by extensities or "organic parts." This distinction holds at different scales: *species* can refer to its usual referent of related groups of organisms, but it could also refer to cell types, whereas *organic parts* can refer to the individual animals that make up a species or to the cells that make up an organ [DeLanda 2002, 54].) Thus the Idea of one species is distinguished from another by changes in differential relations rather than by changes in essential definitions. (I suspect Deleuze would have been delighted by the discovery of the conserved homeobox genes studied in evo-devo [see chapter 10], whose function is precisely to regulate the differential relations of gene expression and protein synthesis, slowing some down [at the limit, preventing them together] and speeding others up.) For Deleuze, it is not that essential definitions or differenciations are impossible; it is just that they occur in a different ontological register from virtual differentiations. Essential definitions belong to the actual register, or the level of properties of formed substances, while Deleuze wants to reach the virtual register, that is, the level of the structures of the intensive processes productive of such actual substances.[14]

Some concrete examples will help us see how this ontology enables us to understand natural processes in multiple registers: physical, meteorological, and social. Deleuze himself often used Gilbert Simondon's (1995) theory of individuation as a very simple model for "actualization." For Simondon, crystallization is a paradigm of individuation: a supersaturated solution is "metastable," and from that preindividuated field—which is "differential" in the sense of being replete with gradients of density that are not crystalline in form but are only "implicit forms" or "potential functions"—crystals are individuated via a process of precipitation. The reason crystallization is only a crude image of other individuation processes is that crystals form in homogenous, albeit differential, solutions, while the Deleuzian virtual involves differential relations among heterogeneous components whose rates of change are connected with each other. For an example of such heterogeneity, let us return to hurricane formation, where it is intuitively clear that there is no central command but a self-organization of multiple processes of air and water movement propelled by temperature and pressure differences. All hurricanes form when intensive processes of wind and ocean currents reach singular points. These singular points, however, are not unique to any one hurricane but are virtual for each actual hurricane, just as the boiling point of water is virtual for each actual pot of tea on the stove. In other words, all hurricanes share the same structure, and that structure (the Carnot cycle) also underlies any heat engine. Finally, in a still more complex social example, Deleuze will interpret Foucault's notion of "discipline" as a multiplicity that allows for the control of any human population. The differential relations here are linkages among rates of change of spatial position, coded movements, complex individual training exercises, and teamwork exercises (Foucault 1977, 167–69). But this multiplicity serves as a "diagram" that guides many different concrete social "assemblages" such as schools, barracks, hospitals, factories, and prisons (Deleuze and Guattari 1987, 530–31n39).

Translated into other terminology, then, Deleuze will say that Ideas or multiplicities are multiply actualizable or "differenciable," but he insists that the underlying structure is virtual or fully differential, that it does not "resemble" the many different concrete systems that actualize it. Using the terminology of Putnam's classic paper "The Nature of Mental States," we can say that functionalism falls prey to Deleuze's resemblance objection because the "Total State" of a system "resembles" its realizations by being fully individuated. Putnam (1975, 434) writes, "A Description of S where S is a system, is any true statement to the effect that S possesses distinct states $S1, S2, \ldots Sn$, which are related to one another and to the motor outputs and sensory inputs by the transition probabilities given in such-and-such a Machine Table. The Machine Table mentioned in the Description will then be called the Functional Organization of S relative to that Description, and the Si such that S is in state Si, at a given time will be called the Total State of S (at the time) relative to that Description."[15] Thus the mental state that can be multiply realized is fully specified or individuated as the "Total State" of the system. It is an individuated pain state (to use Putnam's example), whether it is realized in wetware or hardware, in terrestrial carbon-based life or in some other material. But for Deleuze, an Idea or multiplicity is not individuated or differenciated, even though it is fully differentiated. In this way, the disciplinary Idea, for example, as fully differential or virtual, contains only the relations and singularities into which a human population to be controlled is put. There is nothing prisonlike in the disciplinary Idea: what is put into relation are unspecified populations and unspecified tasks. Thus the elements of the disciplinary Idea are just members of an unspecified population, not prisoners or workers or soldiers or students (these are the components of concrete assemblages), and the relations are merely those of corporeal distance, succession of exercises, precision of movement, degree of obedience to command, and so on, not those of, say, a close-order rifle drill (an example drawn from the concrete military assemblage). The disciplinary Idea can just as well be actualized in a school as in a prison, though (most likely, one hopes) at a different degree of intensity of control.

Doubling Difference

To recap and slightly expand on the preceding discussion, let us turn to the way in which Deleuze's "ontological difference" between virtual multiplicity and intensive individuation processes can help us think the ontological status of "model" in its relation to the events it models. The doubling of difference we can detect in Deleuze's work means that the multiplicity (the model) is differential (a-centered) and that its actualizations are all different events. Translating into other terms, the second claim is that the Deleuzian scheme accounts for the "multiple actualizability" of models.

In *Difference and Repetition* (Deleuze 1994), as we have seen, an Idea or multiplicity is defined as a set of differential elements, differential relations, and singularities. So we here see difference sense 1 ("differentiation"). A multiplicity is "differentiated" (it has no central controlling point) and has the ontological status of "virtuality." Linked intensive processes will individuate or actualize this multiplicity, with a series of cascading determinations occurring when those processes hit singularities or thresholds in their relative values; passing these thresholds triggers qualitative changes in the system. Difference sense 2 ("differenciation") refers to the way in which the multiplicity provides the model of an event type and the different individuated events are separate actualizations of that model. To reinforce this distinction, Deleuze and Guattari (1987, 263–64) provide in *A Thousand Plateaus* a grammar of events and models: the proper name marks the singular event or "haecceity"—"Hurricane Katrina." In this logic, the multiplicity is a pattern and the events are processes conforming to the pattern, thus echoing the title "Difference and Repetition," since although different events are different, they repeat the model, but precisely as different.[16] Note the doubled difference here: (1) the way in which a model is a multiplicity linking many interactive processes (differentiation qua a-centered set) and the fact that (2) for it to be a model, it must be applicable to different events (differenciation qua multiple actualization).

We see how this scheme presents an ontology for the construction of a model for a complex event like a deepwater oil spill. Consider a recent article in *Bioscience*, "A Tale of Two Spills: Novel Science and Policy Implications of an Emerging New Oil Spill Model" (Peterson et al. 2012). With some simplification of the article, we see that a deep-sea oil spill event can be modeled on the basis of a differential multiplicity, that is, the interaction of the following processes: (1) the formation of the oil, gas, and dispersant mixture under the varying conditions of water pressure, water current, turbulence, and so on, at the point of the spill; (2) the buoyancy of oil and gas as affected by water circulation patterns; (3) the formation of droplets of oil by the dispersant's action to create "subsurface oil–gas–water–gas-hydrate emulsions," as these processes relate to the linkage of the acceleration of "microbial degradation" and to the exposure and damage to organisms (such damage itself being the result of multiple processes); and (4) the ripple effects of organism damage throughout food webs; and (5) the effects of the microbially processed oil on food webs, including the seafood people eat.

The important thing for us to note is that each of the links of processes has thresholds at which the system changes its behavior pattern (there is thus a process of progressive determination of the differential field at work in any one oil spill). What makes this a model for future spills is abstracting the relations of the variables from the values of those variables in the Gulf oil disaster. In Deleuzian terms, this moving from the differenciated event token to the differentiated event type is *vice-diction* or *countereffectuation* or *counteractualization*. Now do working scientists need to know the Deleuzian scheme to understand the ontology of models? Of course they do not; the scientists who produced the oil spill model did perfectly well without distinguishing the virtual status of a model from the intensive processes that actualize that model in different events. But that does not mean philosophers should not consider Deleuze's framework as providing an explicit ontology for doubled difference in model and event: the a-centered or differential set of processes in the model and the distinction of differentiated model and differenciated events.

Overview of the Book

In concluding this first section of the introduction, I will only provide brief sketches of the individual chapters, but because each one begins with a forecast of its contents, a more detailed sense of the book's trajectory can be gained by consulting the chapters themselves. Although the occasional nature of the work whence the chapters come prevents any clean narrative overview linking the chapters, I can provide some orienting remarks. In basic overview, the book sets up its theoretical background in this first section of the introduction and then exemplifies it in the second section of the introduction by means of a reading of the work of Francisco Varela in terms of "bodies politic." Then, in the main body of the book, I show the way in which a Deleuzian approach can be articulated with various practices and theoretical reflections: in part I, war and military training; in part II, the 4EA approach to cognitive science; and in part III, new approaches in biology.

Part I has three chapters. In the first chapter, I look to historical research on ancient political economy in the Aegean to sketch a geo-hydro-solar-bio-techno-political multiplicity linking solar, water, and wind energy to the warship and merchant ship; the difference in the daily travel capacity of these ships leads the fifth-century Athenians to empire. In chapter 2, I focus on the way modern military training enables the act of killing in combat in a way that can come back to haunt, psychophysiologically, returning soldiers; the dimensions of the multiplicity here are distance, teamwork, command, and mechanical intermediaries as they intersect the psycho-neuro-physiological makeup of soldiers. Chapter 3 returns to the ancient Eastern Mediterranean and brings together the geopolitical focus of the first with the neurophysiological focus of the second for a fuller account of the multiple suprasubjective (geopolitical), adjunct-subjective (the technical), and subsubjective (neurophysiological) dimensions of the warfare multiplicity.

The four chapters of part II examine the 4EA approach ("embodied, embedded, enactive, extended, affective") to cognitive science. In chapter 4, I develop a Deleuzian-motivated "dynamic interactionist" account of the notion of socially mediated neuroplasticity in Bruce Wexler's (2006) fascinating study *Brain and Culture*. Chapters 5 and 6 are linked treatments of aspects of the "political economy of consciousness." In chapter 5, I examine situations in which the effects of consciousness are attenuated or rendered superfluous in the economy of political action. In chapter 6, on the "granularity problem," I look at the way in which the production of the large-scale patterns of individual consciousness can often be analyzed in terms of subjectification practices that are tied to political economy. Chapter 7, "Adding Deleuze to the Mix," was written as an article for the specialist journal *Phenomenology and the Cognitive Sciences*; it makes the case that Deleuze's ontology can help with questions about the ontological status of perceptual capacities and exercise as well as help deal with the realism–idealism debate.

Finally, the three chapters of part III deal with Deleuze and certain new currents in biology that occupy the intersection among the developmental systems theory of Susan Oyama and colleagues, the enactive approach of Francisco Varela and colleagues, and the eco-devo-evo approach of Mary Jane West-Eberhard. Chapters 8 and 9 prepare the ground for, and then plunge into, the relation of Deleuze's (bio-) panpsychism and the "mind in life" position of Evan Thompson (2007), whereas chapter 10 relates two key concepts of West-Eberhard (2003), unexpressed genetic variation and genetic accommodation, to Deleuze's notions of the virtual and counteractualization.

The main idea of the book is to show how Deleuze's conceptual framework enables us to bring scientifically minded philosophers and philosophically minded scientists—as well as analytic and continental philosophers—into dialogue. There are overlapping themes of affect and of "difference and development," to disabuse us of the customary focus on the rational male adult subject. The book follows *Political Affect* in concentrating on the interplay of the suprasubjective, adjunct-subjective, and subsubjective, following the slogan "above, below, and alongside the subject." That is, in its countereffectuating analyses of differenciated event tokens to reveal the multiplicity or differentiated event types, it moves above the subject to the geopolitical, below the subject to the neurophysiological, and alongside the subject to the social–technical. It similarly follows *Political Affect* in distinguishing three temporal scales, the evolutionary, the developmental, and the behavioral, as well as three compositional scales for bodies politic: the civic, the somatic, and the evental. From the Deleuzian perspective, the patterns, triggers, and thresholds of affective cognitive dispositions are produced via transgenerational subjectification practices that are the intensive individuation processes of a social–neural–somatic multiplicity. Thus the social and the somatic are not synchronic opposites but are linked in a spiraling diachronic interweaving at three temporal scales: the long-term phylogenetic, the mid-term ontogenetic, and the short-term behavioral. The second part of the introduction, to which we now turn, uses the work of Francisco Varela as a way of explaining and exemplifying this conceptual framework.

Varela and Bodies Politic

Francisco Varela's work is a monumental achievement in twentieth-century biological and biophilosophical thought.[1] After his early collaboration in neocybernetics with Humberto Maturana (autopoiesis), Varela made fundamental contributions to immunology (network theory), Artificial Life (cellular automata), cognitive science (enaction), philosophy of mind (neurophenomenology), brain studies (the brainweb), and East–West dialogue (the Mind and Life conferences). In the course of his career, Varela influenced many important collaborators and interlocutors, formed a generation of excellent students, and touched the lives of many with the intensity of his mind, the sharpness of his wit, and the strength of his spirit. In this introduction, I will trace some of the key turning points in Varela's thought, with special focus on the concept of emergence, which was always central to his work, and on questions of politics, which operates at the margins of his thought. I will divide Varela's work into three periods—autopoiesis, enaction, and radical embodiment—each of which is marked by a guiding concept; a specific methodology; a research focus; an inflection in the notion of emergence; and a characteristic political question that specifies a scale of what I will call *political physiology,* that is, the formation of "bodies politic" at the civic, somatic, and evental scales. These terms refer, respectively, to the formation of political states, of politically constituted individuals, and to their intersection in political encounters.

The first period, marked by the concept of autopoiesis, runs from the early 1970s to the early 1980s and uses formal recursive mathematics to deal with synchronic emergence, that is, a focused behavior on the part of an organic system that is achieved via the constraint of the behavior of components of the system; synchronic emergence can be seen as the question of the relation of part and whole. The research focus is on identifying an essence of life. The political question here is the limit of using autopoiesis as a model for enacting social being. Varela sees autopoiesis as only an instance of a general mode of being, organizational closure; he restricts autopoiesis to cellular production—that is, to living systems bound by a physical membrane—and warns against using it as a model of social being. Here we see the question of the macroscale of political physiology, the formation of a body politic in the classical sense, what we will call a *civic body politic*. Varela refuses to countenance the use of autopoiesis as a model for social systems, I will argue, not so much for purely cognitive reasons but because when autopoiesis is enacted, when it is the model for a way of social being, then social systems become obsessed with physical boundaries, leading to a fratricidal zero-sum competition. For him, systems above the cellular level—that is, neurological and immunological systems and social systems—are to be thought as informational systems with organizational closure. (Luhmann, however, will use the term *autopoietic* with regard to those systems as well.) The end result is that autopoietic enactment, in Varela's sense, is solely concerned with synchronic emergence (homeostatic part–whole relations) and is thereby unable to foster the condition for diachronic emergence in social and political dynamics (the emergence of novel patterns from the undoing of former patterns). I will argue that Varela implicitly holds that the historical changes and multiple causation of political systems must be thought in terms of a field whose dynamics are modeled with nonlinear differential equations, which is beyond the scope of autopoietic thought.

The second period, whose concept is that of enaction, spans the late 1980s and early 1990s and uses differential equations to model dynamic systems to deal with diachronic emergence, the production of novel functional structures. The research focus is embodied cognition. In this period, we must distinguish two time scales of diachronic emergence: (1) the fast scale of the coming-into-being of a systematic focus of actual behavior from a repertoire of potential or virtual behaviors and (2) the slow scale of the acquisition of the behavioral modules that form the virtual repertoire available to a system at any one time. The interplay of these scales requires that we think a "virtual self." The political question here is leisure: politics as the system controlling access to training for the acquisition of skills according to the differential access to leisure or free time. Here we see the mesoscale of political physiology, the formation of a somatic body politic as the resolution of the differential relations that structure a dynamic social–political–economic field, a process that is very crudely analogous to crystallization in a "metastable" supersaturated solution.

The third period, whose concept is that of radical embodiment, runs from the mid-1990s to Varela's premature death in 2001 and uses the methodology of neurophenomenology to discuss transversal emergence, the production of distributed and interwoven systems along brain–body–environment lines. The research focus is consciousness (both basic consciousness, or "sentience," and higher-level reflective or self-consciousness) as it arises in the interaction of affect and cognition. With the turn to affect in theorizing concrete consciousness as enacted in distributed and interwoven brain–body–environment systems, we approach the political questions of the other and concrete social perception and hence a microscale of political physiology, the formation of evental bodies politic or, perhaps less barbarically named, political encounters. As we will see, such encounters enfold all levels of political physiology, as a concrete encounter occurs in a short-term social context between embodied subjects formed by long-term social and developmental processes. More precisely—because "context" is too static—a political encounter, like all the emergent functional structures of political physiology, is the resolution of the differential relations of a dynamic field, in this case, one operating at multiple levels: civic, somatic, and evental. (Here we see the limits of the crystallization analogy, as crystals form in homogeneous solutions, while political encounters coalesce in heterogeneous environments.)

Autopoiesis and Synchronic Emergence

Varela is perhaps best known for his early collaboration with Humberto Maturana in developing the concept of autopoiesis. This work, published in Spanish in 1973, and made known to the Anglophone community by a 1974 article and then by a 1980 monograph, is a classic of second-order, or neo-, cybernetics. In our terms, it is marked by a notion of *synchronic emergence,* which is conducted in static part–whole terms. The concept of autopoiesis was developed to provide a horizon of unity for thinking about living entities rather than the haphazard

empiricism of the "list of properties" model usually adopted ("reproduction, metabolism, growth . . ."). In other words, Maturana and
Varela were trying to isolate an essence of life, an essence that would
provide a viewpoint on life that is "history independent" (Varela, Maturana, and Uribe 1974, 187). Varela will come to reject the direct political application of autopoiesis, however.

The Concept of Autopoiesis

To produce the concept of the essence of life, Varela and his colleagues
distinguish organization (essence) and structure (historical accident).
Organization is the set of all possible relationships of the autopoietic
processes of an organism; it thus forms the autopoietic "space" of that
organism (Maturana and Varela 1980, 88; scare quotes in original).
Structure is that selection from the organizational set that is actually at
work at any one moment (Maturana and Varela 1980, xx, 77, 137–38;
see also Hayles 1999, 138; Rudrauf et al. 2003, 31). Changes in the
environment with which the system interacts are known as "perturbations" of the system. The system interacts only with those events
with which it has an "interest" in interacting, that is, those events that
are relevant to its continued maintenance of autopoietic organization (e.g., nutrients). These events of interaction form a process of
"structural coupling" that leads to structural changes in the system.
These changes, as reactions to the perturbation, either reestablish the
baseline state of the system (they reestablish the homeostasis of the
system) or result in the destruction of the system qua living (Maturana
and Varela 1980, 81). Homeostatic restoration thus results in conservation of autopoietic organization. From this essentialist viewpoint, the
origin of life must be a leap into another register, a *metabasis eis allo
genos* ("the establishment of an autopoietic system cannot be a gradual
process; either a system is an autopoietic system or it is not" [Maturana
and Varela 1980, 94]). From the autopoietic perspective, questions of
diachronic emergence have to be thought in terms of "natural drift,"
whose relation to autopoietic essential organization is problematic, as
we will see. In any event, clearly autopoietic organization is synchronic
emergence, in which the whole arises from a "network of interactions
of components" (Varela, Maturana, and Uribe 1974, 187).[2]

The difficulty here is that the assumption of organization as a fixed transcendental or essential identity horizon prevents us from thinking life as the virtual conditions for creative novelty or diachronic emergence. Life for autopoiesis is restricted to maintenance of homeostasis; creative evolutionary change is relegated to structural change under a horizon of conserved organization. If virtual organization is conserved for each organism, no matter the changes in its actual structure—one of the prime tenets of their autopoietic theory—then on an evolutionary time scale, all life has the same organization, which means all life belongs to the same class and has only different structure. As Hayles (1999, 152) puts it, "either organization is conserved and evolutionary change is effaced, or organization changes and autopoiesis is effaced." Autopoietic theory gladly admits all this. "Reproduction and evolution do not enter into the characterization of the living organization" (Maturana and Varela 1980, 96); evolution is the "production of a historical network in which the unities successively produced embody an invariant organization in a changing structure" (104). Although autopoietic theory, developed in the 1970s at the height of the molecular revolution in biology, performed an admirable service in reasserting the need to think at the level of the organism, it is clear that autopoiesis is locked into a framework that posits an identity horizon (organizational conservation) for (structural) change. To summarize, for autopoietic theory, living systems conserve their organization, which means their functioning always restores homeostasis; evolution is merely structural change against this identity horizon.

Let us focus on another key feature of autopoietic systems: the autonomy that they possess in virtue of their synchronic emergence. Their internal complexity is such that coupling with their environment or endogenous fluctuations of their states are only triggers of internally directed action. This means that only those external environmental differences capable of being sensed and made sense of by an autonomous system can be said to exist for that system, can be said to make up the world of that system. The positing of a causal relation between external and internal events is only possible from the perspective of an observer, a system that itself must be capable of sensing and making sense of such events in *its* environment.

Quite soon after writing *Autopoiesis* with Maturana in 1973, Varela came to restrict the validity of the idea of autopoiesis to the cellular level, rejecting the use of autopoiesis as a concept for thinking social systems. In this period of his work, Varela distinguishes between autopoiesis, which is limited to physical production within the spatial border provided by a cellular membrane, and organizational closure, which can be applied to systems with an informational component. Varela thus comes to insist on the complementarity of two forms of explanation: autonomy versus control or, what amounts to the same distinction, autopoietic versus informational-symbolic explanations. In "On Being Autonomous: The Lessons of Natural History for Systems Theory," Varela (1977) insists that autonomy and control perspectives are complementary. At this period of his work, Varela is working with a recursion model of closure, in which the "closure thesis" states that "every autonomous system is organizationally closed" and organizational closure entails the "indefinite recursion of component interaction" (79). Here Varela distinguishes cells as "physically independent units" from "systems where autonomy is expressed in an 'informational' way . . . nervous and the immune system of animals, which are, as it were, cognitive systems in the macroscopic and microscopic domains of the organism" (79). It is this distinction between physical production enclosed in a physical space and the "information" of distributed systems that will lead him to restrict autopoiesis to the cellular level. "Information," of course, must be in scare quotes as the cognition Varela is talking about entails structural coupling and triggering of autonomous response rather than recovery of objective information.

Here Varela posits limits of "differentiable dynamic representation" (modeling of the changes in systems) due to the limited ability at the time to handle the differential equations necessary to model nonlinear dynamic systems (81) and so opts for his self-referring, indefinite recursion model, which needs "an infinite-valued logic" (82). Operator trees are constructed, and "circularity is captured through the solutions or eigenbehaviors of equations in this operator domain." This allows a "representation of autonomy which is not so abstract as indicational forms, and yet not so demanding of quantitative detail as in differentiable dynamic descriptions" (82). The paper closes with a clear statement of Varela's constructivism and antirealism: "the contents of our reality are truly a reflection of the recursive biological and cognitive computations . . . there is more a construction than a map"

(82). We will see how what Varela would call the *autopoietic* enactment of this autonomous constructivism, whereby a system comes to focus on what it is already set up to see as being in its "interest" in maintaining its *physical* boundaries, will have disastrous effects when such an "epistemology" is instantiated in a political system producing mutually blind—and hence fratricidal—competing systems, in a time of civil war.[3]

In the meantime, we should stick with the question of modeling of systems. In *Principles of Biological Autonomy*, Varela (1979a) explains that he is attracted to dynamic systems models but finds them limited to the molecular level and suggests algebraic–formal recursion models as the most general kind to use in modeling larger systems. "The classical notion of stability in differentiable dynamics is the only well-understood and accepted way of representing autonomous properties of systems. . . . [We can find] excellent examples of the fertility of this approach for the case of molecular self-organization" (203). However, this approach has a restricted validity: "An underlying assumption, is, however, that there is a collection of interdependent variables, and it is the reciprocal interaction of these component variables that brings about the emergence of an autonomous unit. . . . [Thus] the differentiable dynamic description becomes a specific case of organizational closure" (203). More precisely, the dynamic systems approach is of limited validity for organisms (and political systems, as we will see), among which we find a number of interlocking and embedded informational or symbolic systems: "At the same time one finds the limitations imposed by [the differentiable framework]: More often than not, autonomous systems cannot be represented with differentiable dynamics, since the relevant processes are not amenable to that treatment. This is typical for informational processes of many different kinds, where an algebraic–algorithmic description has proven more adequate. Accordingly, the fertility of the differentiable representation of autonomy and organizational closure is mostly restricted to the molecular level of self-organization" (203).

The difference between the dynamic and the formal models depends on the difference between an abstract temporal approach and a concrete spatial approach. Varela (1979a) refers to the dynamic approach of Eigen and Goodwin as that which focuses on a "network of reactions and their temporal invariances, but disregard on purpose the way in which these reactions do or do not constitute a unit in space" (204). In this emphasis on physical boundaries and material production, we see what leads Varela once again to insist on the need for complementarity between control and autonomy perspectives in which dissipative systems are treated as input–output fluxes. Although he claims that there is some evidence of dynamic models being able to capture membrane formation, as in Zhabotinsky reactions, "it is still a matter of investigation how well the differentiable-dynamics approach can accommodate, in a useful way, the spatial *and* the dynamic view of a system" (204). Beyond the molecular level, we reach our cognitive limits, set by the state of knowledge at the time: "But it is in going beyond the molecular level, where we cannot rely on a strong physico-chemical background of knowledge, that the insufficiency of the differentiable framework appears, and thus the need to have a more explicit view of the autonomy/control complementarity, and an extension of differentiable descriptions to operational/algebraic ones" (205).

In other words, at the time of *Principles,* Varela thought that cellular autopoiesis could be thought dynamically and that, while neurological and immunological processes are "borderline cases" (205), higher-level processes, organismic and social, could not be.[4] The key question is the ability to represent metastable (changeable, creative) systems. That was impossible in 1979 with the algebraic approach; we are left with a series of questions for further research:

> Clearly, both approaches cover somewhat *non*-overlapping aspects of systemic descriptions. Thus, it is necessary to have a way of dealing with plasticity and adaptation. Natural systems are under a constant barrage of perturbations, and they will undergo changes in their structure and eigenbehavior as a consequence of them. There is no obvious way of representing this fundamental time-dependent feature of system–environment interactions in the present algebraic framework. In contrast, the question of plasticity is a most natural one in differential frameworks because of the topological properties underlying this form of representation: hence the notions of homeorrhesis and structural stability in all their varieties. To what extent can the experience gained in the

differentiable approach be generalized? How can notions such as self-organization and multilevel coordination be made more explicit in this context? Is category theory a more adequate language to ask these questions? These and many more are open questions.[5] (205–6)

Politics and Autopoiesis

The political question in this first period is the extension of autopoiesis as a model for enacting social being, the question of the body politic in its classic sense—what we call the macroscale of political physiology. Varela will reject all attempts at such an extension. The tension with Maturana on this point is evident in the 1980 English publication of *Autopoiesis*, in which the authors note that they are unable to agree "on an answer to the question posed by the biological nature of human societies from the vantage point" of autopoiesis (Maturana and Varela 1980, 118). Varela's departure from Maturana is apparent in "On Being Autonomous," in which autopoiesis is said to suggest a "universal feature" shared by many other types of systems, to wit, "organizational closure," which extends beyond physical systems to "informational" systems (Varela 1977, 79). In "Describing the Logic of the Living," Varela (1981, 37; emphasis original) is crystal clear: "autopoiesis is a particular case of a larger class of organizations that can be called *organizationally* closed, that is, defined through indefinite recursion of component relations." After insisting on some concrete sense of "production" to define autopoiesis, Varela drives home his point: "Frankly, I do not see how the definition of autopoiesis can be *directly* transposed to a variety of other situations, social systems for example" (38; emphasis original).

In a late interview, "Autopoïese et émergence," Varela (2002, 170; my translation) gives his reasons for resisting an extension of autopoiesis to the social:

It's a question on which I have reflected for a long time and hesitated very much. But I have finally come to the conclusion that all extension of biological models to the social level is to be avoided. I am absolutely against all extensions of autopoiesis, and also

against the move to think society according to models of emergence, even though, in a certain sense, you're not wrong in thinking things like that, but it is an extremely delicate passage. I refuse
to apply autopoiesis to the social plane. That might surprise you,
but I do so for political reasons. History has shown that biological
holism is very interesting and has produced great things, but it has
always had its dark side, a black side, each time it's allowed itself
to be applied to a social model. There's always slippages toward
fascism, toward authoritarian impositions, eugenics, and so on.

What is the key to the "extremely delicate passage" necessary to think
social emergence while avoiding the "dark side" of the slide into fascism? First, we should note the complete rejection of autopoietic
social notions, while the notion of social emergence is less strongly
condemned. I would argue that the difference lies in Varela's concep-
tion of autopoiesis as synchronically emergent, which locks out the
sort of diachronic emergence we will study in the next section. If
one could think the formation of civic bodies politic using dynamic
systems modeling (something that for Varela, at the time of *Principles*,
was considered impossible, as we have seen), if one could see them
as resolutions of the differential relations inherent in a dynamic field
(again, something crudely analogous to crystallization in supersaturated solutions or lightning as the resolution of electric potential differences in clouds or weather systems as resolution of temperature
differences in air and water), then we would at least have the possibility
of an "extremely delicate passage" in thinking political change. But
without that possibility of novel production, modeled by dynamic
systems means, autopoietic social systems, once formed and mature,
construct a world only in their own image and, when locked in conflict
with another such system, cannot ascend to an "observer" status that
would see them both as parts of a larger social system. Instead, the
two conflicting systems are locked in fratricidal combat, producing a
torn civic body politic, producing civil war.

Let us turn here to "Reflections on the Chilean Civil War" for some
historical detail about Varela's worries about the political misuse of
"biological holism," or a misapplication of autopoiesis in enacting the
macroscale of political physiology, the formation of a society or body

politic. In this discussion, "epistemology" is not a matter of neutral understanding but of enactment, of the bringing into being of a way of social living. The stakes are the highest possible for Varela in this deeply personal and emotional piece: "epistemology does matter. As far as I'm concerned, that civil war was caused by a wrong epistemology. It cost my friends their lives, their torture, and the same for 80,000 or so people unknown to me" (19). Varela's analysis shows that Chile had become polarized into two separate worlds without communication, that is, one could claim, two "autopoietic" systems with no sense-making overlap, no means of mutual recognition, but only a concern with physical boundary maintenance: "the polarity created a continual exaggeration of the sense of boundary and territoriality: 'This is ours; get out of here'" (16). I read this as Varela indicating the dangers of extending autopoietic notions to the social. The danger lies not in using autopoiesis as a means of understanding the social but in using autopoiesis as a model in enacting a way of social being. An autopoietic social being is one focused on boundary maintenance, and this focus can create a fratricidal polarity.

The key to understanding Varela's prohibition on extending autopoiesis to social systems, that is, his move "beyond autopoiesis"—but not beyond neocybernetics as concerned with organizational closure of informational systems—is to appreciate his warning against enacting the concern with physical boundary protection, which carries along with it the risk of falling into "polarization." Varela recounts his moment of insight when he overcame that polarization: "polarity wasn't anymore this or that side, but something that we had collectively constructed"; political worlds, previously autonomous, had to be considered merely "fragments that constituted this whole" (18). The problem, of course, is establishing the "observer" position, which can use the notion of the interaction of organizationally closed informational systems to appreciate this larger whole encompassing the autonomous and mutually blind systems. Varela finds this position in Buddhist practice, with its necessity of stressing the "connection between the world view, political action and personal transformation" (19). To avoid the fratricidal polarization of competing autonomous systems, relativistic fallibility is the key to the construction of a political world: "we must incorporate in the *enactment*, in the projecting out of our world views, at the same time the sense in which that projection is only one perspective, that it is a relative frame, that it must contain a way to undo itself" (19; emphasis added). Such flexibility, as we will see next, is available to a system producing a "virtual self" out of a multiplicity of coping resources, out of a repertoire of behaviors, but is foreclosed to the physical cellular systems to which Varela consigns autopoiesis. For that reason, the autopoietic model of cellular systems is disastrously misapplied when it is used to *enact* the macroscale of political physiology, as in the brutally violent "epistemology" (qua way of social being) enacted by the conflicting sides in the Chilean Civil War. To summarize Varela's position: *enacting* autopoiesis as a way of social being (as distinguished from using the concept of organizationally closed informational systems to *understand* a social situation) turns a social field into a polarized confrontation of systems seeking physical boundary maintenance; focused on synchronic emergence or part–whole relations, which it sees in zero-sum terms ("this is ours; get out of here"), such autopoietic enactment cannot foster the conditions for the diachronic emergence of historical novelty.

The Virtual Self and Diachronic Emergence

With this invocation of the key term *enactment*, we can move to the second period of Varela's work, the late 1980s and early 1990s, in which the recursive models of systems Varela used in *Principles* under the acknowledged influence of Spencer-Brown's *Laws of Form* drop away as dynamic systems modeling makes progress, especially in connectionist work in cognitive science. Here we see that Varela's work develops a notion of *diachronic emergence* (emergence as the production of novel structures).[6] In this period, Varela broke into his own with a series of fundamental works on Artificial Life, immunology, and the status of the organism. This period culminates with his second most well-known work, *The Embodied Mind* (Varela, Thompson, and Rosch 1991), the manifesto of the enactive school in cognitive science.[7]

In this second period of his work, Varela deals with three cognitive registers: immunological, neurological, and organismic (which includes the previous two). We will concentrate on the intersection of the neurological and the organismic but should not forget Varela's groundbreaking work theorizing the immune system as a network, which rejects the military metaphor of protection of interiority and resolves the paradoxes of self versus nonself recognition that beset the classic concept (see, e.g., Varela and Coutinho 1991).

The Enactment of Cognition

The inflection of emergence in the period of enactment or the virtual self is diachronic emergence, which operates at two temporal scales in both neurological and organismic registers. On the fast scale in the neurological register, we find resonant cell assemblies, which arise from chaotic firing patterns; on the fast scale in the organismic register, we see the arising of behavioral modules, or *micro-identities*, from a competition among competing modules. We can see that both these modes of diachronic emergence on the fast scale are resolutions of a dynamic, metastable differential field. Whereas Varela concentrates on the fast scale, we will examine the slow scale, the acquisition of behavioral modules in those registers, for here we intersect the political question of leisure and access to training for acquisition of skills. The differential field here is the field of formation of "somatic bodies politic," the mesoscale of "political physiology."

Working from connectionist models, but rejecting their representationalist assumptions, Varela looks to resonant cell assemblies (RCA) as the neurological correlate for micro-identities. The latter concept comes from phenomenological reflection on the concrete life of the everyday. Following Heidegger and Merleau-Ponty in opposing a Cartesian heritage privileging self-conscious, reflective, and verbal reasoning as the essence of cognition, Varela will claim that most everyday life (of competent adults, to be sure) is accomplished in skilled, nonreflective comportments. Disruptive social encounters, however, lead to breakdowns in such everyday coping and can lead to reflective decision making or to the adoption of another skilled comportment (Varela 1991; 1992a). The neurological correlates of breakdowns are a fall into a background of chaotic firing, out of which emerges a new RCA. This resolution of the differential field of widely distributed chaotic firing forms the basis for creativity in the arising within the organism of a triumphantly emergent comportment. There is no choice here, as the process of arising of an RCA is too fast for conscious reflection, which occurs in temporal chunks, so that RCA formation occurs behind the back of reflective consciousness. An RCA is the neurological correlate of what is described in other registers as a skill or agent or module, and the creative emergence occurs on the basis of the historical formation of a repertoire of behavioral modules.

We see here two important concepts: the virtual self and the enactment of world. As this repertoire is a distributed and modular system, both at the behavioral and the neurological level, Varela will talk of a "virtual self" or "meshwork of selfless selves," as the subtitle of Varela (1991) puts it. The correlate of the virtual self, with its multiplicity of micro-identities, is the enacted world. The laws of physics, or the regularities of the environment (the epistemological niceties that might distinguish these phrases need not concern us here), form only loose constraints for the worlds each organism brings forth or enacts in a process of "surplus signification." Here we see echoes of the sense-making at the heart of the autopoietic notion of "structural coupling," but with more ability to flesh out the neurological processes at work.

With these two concepts, as well as Heidegger and Merleau-Ponty, in mind, Varela, Thompson, and Rosch (1991, 150) write in *The Embodied Mind*,

> The challenge posed by cognitive science to the Continental discussions, then, is to link the study of human experience as culturally embodied with the study of human cognition in neuroscience, linguistics, and cognitive psychology. In contrast, the challenge posed to cognitive science is to question one of the most entrenched assumptions of our scientific heritage—that the world is independent of the knower. If we are forced to admit that cognition cannot be properly understood without common sense, and that common sense is none other than our bodily and social history, then the inevitable conclusion is that knower and known, mind and world, stand in relation to each other through mutual specification or dependent coorigination.

A Challenge to Enacting Politics

At this point, I would like to shift from exposition to critical engagement by extending this series of challenges so that enaction is in turn challenged to examine the unconscious social grouping hiding in the "our" of "our bodily and social history." The challenge is to examine the historical and political system that distributes leisure and the access to training for learning of behavioral modules. A further challenge is to disabuse ourselves of the naive notion that all those modules are beneficial to the body that incorporates them rather than beneficial to the power structure of the society. In other words, many people incorporate behavioral modules that hurt them, although those modules help reproduce inequitable social dynamics.[8]

We see the contours of this problematic in works of the 1990s. The "constitution" of the "cognitive agent" is "a matter of common-sensical emergence of an appropriate stance from the entire history of the agent's life. . . . The key to autonomy is that a living system finds its way into the next moment by acting appropriately out of its own resources. And it is the breakdowns, the hinges that articulate microworlds, that are the source of the autonomous and creative side of living cognition" (Varela 1992a, 11). Once again, we have to distinguish two temporal scales of diachronic emergence: "the moment of negotiation and emergence when one of the many potential microworlds takes the lead . . . the very moment of being-there when something concrete and specific shows up. . . . Within the gap during a breakdown there is a rich *dynamic* involving concurrent subidentities and agents" (Varela 1992a, 49). This is the fast dynamic. If we are to critically engage Varela's work, we also need to thematize how the behavioral repertoire that provides the scope of those many potential microworlds has emerged over the slow scale of development, maturation, and learning. In other words, we must think of the slow dynamic of structural coupling leading to the ontogenesis of the embodied subject, a process that must be analyzed politically as the differential access to training. To bring out all its potential, Varela's insistence on autonomous organisms needs to be supplemented with an analysis, using social–political categories, of the distribution of access to training that allows differential installation of modules–agents–skills in a population of organisms.

The important thing is not to confuse autonomy and competence. A corporeal subject with a limited repertoire of capacities, or with a repertoire of disempowering habits, is still autonomous in the Varelean sense, as producing behaviors on the basis of environmental triggers or endogenous fluctuation. No matter how wide or narrow your repertoire of skills, no matter how powerful or weak you are in enacting them, you are no more autonomous than is any other organism in any one action. However, there is a difference in competence, how well your actions enhance your survival and flourishing and those of others, as well as a difference in the range of environmental differences you can engage and survive, thus preserving your autonomy for future encounters. But you have to be trained to acquire many of these skills. As Varela (1999a, 17) puts it in *Ethical Know-How,* "the world we know is not pre-given; it is, rather, enacted through our history of structural coupling, and the temporal hinges that articulate enaction are rooted in the number of alternative microworlds that are activated in each situation." Again, to develop more fully Varela's insight and thus to reach the full concrete reality of our social life, we have to analyze politically that history of structural coupling in terms of access to training to greater or lesser number and greater or lesser quality of skills opening microworlds.

The key to thematizing this mesoscale of political physiology is to think of downward causation in social emergence, the macroscale of the body politic to which we referred earlier. Picking up here on a contemporaneous essay written with Jean-Pierre Dupuy (Dupuy and Varela 1992), Varela (1999a, 62) describes in *Ethical Know-How* the way upward causality allows for the emergence of social regularities: "interactions with others . . . out of these articulations come the emergent properties of social life for which the selfless 'I' is the basic component. Thus whenever we find regularities such as laws or social roles and conceive them as externally given, we have succumbed to the fallacy of attributing substantial identity to what is really an emergent property of a complex, distributed process mediated by social interactions." But here Varela is working with a formal model of synchronic emergence and has neglected the downward causality of these regularities, whether institutionalized in disciplinary intervention or distributed as modulating "control," as they work in the slow temporal scale of the diachronic formation of somatic bodies politic in the context of a particular constellation of a civic body politic. As generations go by, we see a patterned differential social field, channeling perception, action, and affect, along lines of social roles. Varela has only demonstrated that laws, rules, institutions, and so on, are emergently produced by upward causality in a synchronic emergence; he has neglected to show the downward causality effected by these regularities (which we could model by tracking the formation of attractors in a social space representing social habits) and the way this socially enacted world structurally couples with, and guides, the ontogeny of the individual person. It's the prepersonal social field that needs to be thought, as persons are resolutions of the differential social field, concretions that form the affective topology of the person: the patterns, thresholds, and triggers of basic emotions or affective modules of fear, rage, joy, and so on, as they interact with the cognitive topology of the person, the cognitive modules or basic coping behaviors that make up the everyday repertoire of the person.

Neurophenomenology and Transverse Emergence

In developing the practice of *neurophenomenology,* a concept he produced, Varela (1996) begins his late period. It is in this period that the point of contact with politics appears in the question of concrete and affective social perception, the formation of the evental body politic or the political encounter, what we will call *transverse emergence.* This latter term indicates the formation of a functional structure involving organic systems and environmental objects, including technological items, as we see in "extended cognition" involving the use of physical marks, ranging from simple scratches in clay tablets to calculators, computers, and the like.[9]

Interlocking the Neural, Somatic, and Environmental

In a late and very important article, collaborating for the last time with Evan Thompson, Varela writes, "Neural, somatic and environmental elements are likely to interact to produce (via emergence as upward causation) global organism–environment processes, which in turn affect (via downward causation) their constituent elements" (Thompson and Varela 2001, 424). There is a slight terminological nuance here, as Varela has always distinguished *environment* (as objectivist or realist) from *world* (as enactive). We are to read this distinction as maintaining that the environment (= laws of nature or physical regularities) provides constraints on world making, but constraints only, and doesn't optimally specify those worlds. Thus, to use the classic example from *The Embodied Mind*, light obeys laws of physics, but that only provides constraints on the construction of many different enacted color-worlds, which track lines of natural drift. The precision is that we do not see structural coupling between organism and world but between organism and environment, with the latter coupling being the process of the enactment of world. With this in mind, we note that Thompson and Varela specify three dimensions of "radical embodiment":

1. Organismic regulation, in which affect appears as a "dimension of organismic regulation . . . the feeling of being alive . . . inescapable affective background of every conscious state"

2. Sensorimotor coupling, in which "transient neural assemblies mediate the coordination of sensory and motor surfaces, and sensorimotor coupling with the environment constrains and modulates this neural dynamics. It is this cycle that enables the organism to be a situated agent"; insofar as "situated agent" means "that which enacts a world," we see that coupling with the environment constrains and enables world making

3. Intersubjective interaction, in which "the signaling of affective state and sensorimotor coupling play a huge role in social cognition. . . . Higher primates excel at interpreting others as psychological subjects on the basis of their bodily presence (facial expressions, posture, vocalizations, etc). . . . Intersubjectivity involves distinct forms of sensorimotor coupling, as seen in the so-called 'mirror neurons' discovered in area F5 of the premotor cortex in monkeys. . . . There is evidence for a mirror-neuron system for gesture recognition in humans, and it has been proposed that this system might be part of the neural basis for the development of language" (424)

We should note here that the thought of intersubjectivity in Varela's late period stems from the notion of "the other" as developed in the theory of the recognition of the alter ego, based on Husserl's Fifth Cartesian Meditation (although supplemented by the recognition of recent research into mirror neurons). For example, Varela (1999c, 81) writes in a popularizing article,

> It is best to focus on the *bodily* correlates of affect, which appear . . . as directly felt, as part of our *lived body*. . . . This trait . . . plays a decisive role in the manner in which I apprehend the other, not as a thing but as another subjectivity similar to mine, as alter ego. It is through his/her body that I am linked to the other, first as an organism similar to mine, but also perceived as an embodied presence, site and means of an experiential field. This double dimension of the body (organic/lived; *Körper/Leib*) is part and parcel of empathy, the royal means of access to social conscious life, beyond the simple interaction, as fundamental intersubjectivity.[10]

The "Other" and Concrete Political Encounters

To see how the problematic of the "other" is an abstract "philosopher's problem," let us note that in *The Embodied Mind*, Varela and his collaborators, Evan Thompson and Eleanor Rosch, cite Rosch's research into categorization, in which, in Rosch (1978), she poses a "basic level" of perception–action–linguistic naming in a hierarchy of abstraction. This basic level is, in her example, "chair" rather than "furniture" or "Queen Anne." In the same article, Rosch proposes a "prototype" theory for internal category structure—rather than an ideal exemplar, we have concrete prototypes by which we judge category membership by how close or far an object is to our prototype, not by whether it satisfies a list of necessary and sufficient conditions that we carry around with us. If we adopt Rosch's model, in concrete social perception, we are never faced with the Husserlian problem "is this just a thing or is it an alter ego?" which we resolve by distinguishing between things and subjects. Rather, we are always confronted with other people at basic-level social categories appropriate to our culture: for us today, the famous age, size, gender, race, class system. So we never see another subject; instead, we see over here a middle-aged, small, neat, fit, professional black woman (say, Condoleezza Rice) or an elderly, patrician, tall white man (say, George H. W. Bush). So we have to say that Varela's discussion in "Steps" is abstract, which is revealed by his use of "his/her." In our society, we never *perceive* a subject we can call "his/her": we can posit such a creature, but that's a refined political act of overcoming our immediate categorization process, by which we perceive gendered subjects, to construct an abstraction we can call a nongendered "intersubjective community" or "humanity" or some such. While this might be a worthy ethical ideal for which we can strive, it's just simply not what we perceive "at first glance."

It is not that we are completely without guidance here regarding social perception. In their "At the Source of Time" article, Varela and Depraz (2005, 68) mention what would need to be fleshed out: among the five components of affect, the first is "a precipitating *event,* or trigger that can be perceptual (a social event, threat, or affective expression of another in social context) or imaginary (a thought, memory, fantasy . . .) or both." In other words, the social trigger has to recognizable, based on the ontogeny of the perceiving subject. As we claim earlier, this ontogeny has to be thought as a resolution of a prepersonal dynamic differential social field. After learning our mid-level social categories,[11] we never immediately encounter an "other," only concrete people we locate in a complex social categorization scheme. The encounter with the "other" is the result of an abstraction, a working up of the initial encounter, abstracting away from the mid-level categories of concrete social perception.[12]

Let us conclude by returning to "Reflections on the Chilean Civil War," in which Varela (1979b, 18) provides an example of mid-level categories in concrete social perception and affect: "I remember very well that the soldier, whom I saw machinegunning the other fellow who was running down the street, was probably a 19-year old boy from somewhere in the South. A typical face of the people of the South . . . I could see in his face what I had never seen, a strange combination of fear and power."

Varela's reminiscence rings true to concrete social perception. He didn't see a neutral subject, an "other"; he saw a southern Chilean boy of nineteen, a concrete person who is gendered, aged, and racially or at least ethnically marked. In that marking, and in the perception of a new affective state on the soldier boy's face, that "strange combination of fear and power," we engage all scales of political physiology: the macroscale of a civic body politic torn apart in civil war; the mesoscale of the development of the repertoire of behavioral modules, as the boy is marked by this affective combination; and the microscale of political encounter, mediated by affect and cognition on Varela's part as this assemblage or momentary transversal emergence arises—street, gun, soldier, shooting, running, dying, observing. Our challenge is to negotiate the "extremely delicate passage" of social emergence that would let us think through the interchanges of all levels of political physiology in this haunting scene, civic, somatic, and evental at once.

PART I

Geophilosophy: Earth and War

≈ 1 ≈

Geo-hydro-solar-bio-techno-politics

I hope the barbarous nature of my title will be forgiven once it is realized that earth, wind, water, and sun are not just the elements of mythology but are also dimensions of internally differentiated multiplicities whose individuations interlock them with weapons, tactics, and social, political, and physiological processes to produce differenciated "bodies politic." That is to say, using the language of *Political Affect,* in this chapter, I enter into the nonhuman elements implicit in the construction of bodies politic as imbrications of the social and the somatic. In still other words, the "social" is just as much an interlocking of the natural and cultural as is the "somatic." Here I first look to *geohistory* and the ontology and history of *hydropolitics,* then bring them together with *bio-solar-politics* and *political physiology.* In this way, I look to decenter, or better, to embed, military and political subjects by looking to emergent processes above, below, and alongside subjects. I look above to the geopolitics of circuits of food qua captured solar energy, below to political physiology qua entrainment-provoked solidarity, and alongside to biotechnical assemblages such as the phalanx and the trireme. As we will see, in this chapter, there is a bit more focus on the suprasubjective geopolitical and on the adjunct-subjective technical than on subsubjective political physiology as such, whereas in chapter 2, those emphases are reversed. Chapter 3, the final one on warfare, will be more closely balanced among the supra-, adjunct-, and subsubjective.

Geohistory

It seems that Deleuze and Guattari prefer geography to history in *A Thousand Plateaus.* Do they not come right out and say that "nomads have no history; they only have a geography" (Deleuze and Guattari 1987, 393)? Like many of their oracular sayings (e.g., "God is a lobster" [40]), this one makes perfect sense within their framework. In exploring this slogan, we can say that geography has three registers for them, corresponding to their three ontological registers of actual, intensive, and virtual. There is *actual geography*, the production of maps (or, better, *tracings*) of extended space, most often in the service of the State and its traditionally primary mode of spatialization, striation (12); *intensive geography*, the cartography that follows lines of becoming, the mapping of longitudes and latitudes of bodies (260–61); and *virtual geography*, the map of the abstract machines of coding, overcoding, and decoding as they are actualized in the machinic assemblages of tribes, empires, and war machines (222).

Before we go any further, we should note that there is much more to say about Deleuze and Guattari's discussion of spacing practices (smoothing and striating) than we will treat here. For instance, there is the imbrication of spacing practices and the numeration practices (numbering and numbered number) as they differ in nomad and State social machines (380–94). We should note also that the smoothing of spaces by the armed forces and surveillance technologies of contemporary States and quasi-State corporations—as well as the demand for identity cards and fixed addresses by asylum seekers, economic migrants, and other displaced persons—prevents any naive postmodernist privileging of smoothing as somehow progressive or always beneficial and striating as always retrograde and repressive. In this vein, Weizman (2007) treats the practical application by the Israel Defense Force of Deleuze and Guattari's theories of the smoothing and striating of space in operations in Palestinian urban space, while Hardt and Negri (2000) call attention to the desire for stability and identity on the part of many displaced persons today. With all that in mind, we do well to remember that the main text of *A Thousand Plateaus* ends with the line, "Never believe that a smooth space will suffice to save us" (Deleuze and Guattari 1987, 500).

To return to our simplified sketch, history, by contrast to geography, is the time-keeping and self-fulfilling prophecy of States. It is the constructed sequence of significant events that seemingly obliterates geography, the earth, and the nonhistoric presignifying ("primitive" [211–12]) and countersignifying ("nomad" [118]) regimes: "the defeat of the nomads was such, so complete, that history is one with the triumph of States" (394). Deleuze and Guattari's view of official history is as hostile as their view of geography is sympathetic. They take for granted that history is a fabrication by sedentary societies to justify their importance and centrality. Even at its best, "all history does is translate a coexistence of becomings into a succession" (430); that is, history establishes conditions for translation of coexisting diagrams, for the actualization of virtual multiplicities. In this way, "the task of the historian is to designate the 'period' of coexistence or simultaneity of these two movements" (deterritorialization and reterritorialization) (221; for an extended scholarly account of Deleuze and Guattari's philosophy of history, see Lampert [2006]).

As always with Deleuze and Guattari, however, the de jure opposition of geography versus history is matched by the recognition of their de facto mixing. Thus, in the section on geophilosophy in *What Is Philosophy?* (Deleuze and Guattari 1994), we see Deleuze and Guattari, in affirming the necessity of a "geophilosophy," also affirm the necessity of a "geohistory," following Braudel: a materialist history sensitive to the earth, a historical geography or geographical history (95). Now in thinking about the *geo-* of *geohistory*, we have to recognize first of all that the French word *terre* in *A Thousand Plateaus* has various meanings that interweave ontology and politics in what I have elsewhere called *political physics* (Protevi 2001). *Terre* has at least four registers, the first three of which are equivalent to the English "earth" and the fourth to the English "land"[1] (Deleuze and Guattari 1987). In *A Thousand Plateaus, earth* is (1) equivalent to the virtual plane of consistency on which strata are imposed (Deleuze and Guattari 1987, 40); (2) part of the earth–territory *(terre–territoire)* system of romanticism, the gathering point, outside all territories, of "forces of the earth" for intensive territorial assemblages (338–39); and (3) the "new earth" *(une nouvelle terre)*, the correlate of absolute deterritorialization, tapping "cosmic forces" or new potentials for creation (423; 509–10). *Land*, by contrast, is *terre* that is constituted by the overcoding of territories under the signifying regime and the State apparatus (440–41).

When we look for examples of Deleuze and Guattari's interest in geohistory, we find, among the more interesting references in *A Thousand Plateaus*, Karl Wittfogel's (1957) Cold War epic *Oriental Despotism*. Wittfogel claimed that a particular form of social organization, "Oriental despotism"—characterized in its most intense form by the

State being the largest landowner; by the bureaucratic organization of landlords, capitalism, and the gentry; and by a form of "weak private property" allowing for "total power" rooted in the State—came from State origination and control of massive irrigation projects. Deleuze and Guattari refer to Wittfogel at several points in *Anti-Oedipus* (Deleuze and Guattari 1984, 211, 220) and *A Thousand Plateaus* (Deleuze and Guattari 1987, 19, 363) concerning the State as apparatus of capture. We cannot take Wittfogel at face value, for we can see that his primary libidinal political investment was anticommunist rather than anti-"despotic"; he left behind his geohistorical focus when he tried to show that Stalinist totalitarianism could be traced to cultural influences via some unspecified sort of contact with Oriental despotism. Although himself a victim of Nazism, spending nine months of 1933 in concentration camps before being freed as the result of intense negotiation and family contacts (Ulmen 1978, 162–69), he never traced the "total power" of the Nazi regime (which certainly had no connection with irrigation), nor did he investigate American West waterworks, which, although admittedly not "total" power, nonetheless involve power and stratification (Worster 1985, 28–29). In any case, although it is not the single determining factor, Wittfogel does point us toward an important aspect of geohistory: the management of water, leading to what we can call hydro-bio-politics. To see the contours of this complex arena, let us first trace the ontology and politics of water.

Hydro-ontology

For Deleuze and Guattari, being is production. As we have seen in "Introduction I," the production process (intensive difference driving material flows resulting in actual or extensive forms) is structured by virtual multiplicities. Multiplicities are composed of mutually defined elements with linked rates of change ("differential relations") peppered with singularities. In the mathematical modeling of physical systems, singularities are points at which the graph of a function changes direction. Singularities in models represent thresholds in intensive processes, where a system undergoes a qualitative change of behavior. Being as production is symbolized in *Difference and Repetition* by the slogan "the world is an egg" (Deleuze 1994, 251). What this means is that "spatiotemporal dynamisms" or intensive individuation processes actualize or differenciate Ideas. These intensive processes, however, are hidden by the constituted qualities and extensities of actual products. The example of embryology shows this doubled difference, or differenciation of differentiation, as the dynamic of the egg's morphogenesis implies a virtual multiplicity of genetic and epigenetic factors incarnated in such a way that there are things only an embryo can do or withstand, actions of folding and so on that are impossible for an adult. In other words, individuation processes are progressive determinations of a virtual differentiated multiplicity on the way to an actual differenciated entity. Philosophical thought, however, is vice-diction or countereffectuation: it goes the other way from production. Thinking philosophically is a matter of establishing the multiplicity of an entity, event, or process—"constructing a concept" in the words of Deleuze and Guattari (1994)—by moving from extensity through intensity to virtuality.

Following water is a great way to think in the Deleuzian manner. First, we can measure water *extensively* in its three forms, solid, liquid, gas: the thickness in feet of the Greenland ice shelf; the number of cubic kilometers of ocean water; the percentage of humidity at one time and place, and so on. Second, we can follow water flows as *intensive* processes. Differences in temperature, density, and so on, provoke material flows, for example, the manifold ocean currents: to name only a very important one, the Gulf Stream brings equatorial heat north, warming northwestern Europe and sinking off Greenland as temperature drops, density and salination increase, and the stream plummets to the ocean floor to join the subsurface ocean currents. Third, we can construct the *virtual* Idea of water. Among its dimensions would be those that govern the hydrological cycle: linked rates of change of difference-driven intensive processes of evaporation, precipitation, and runoff with singularities marking phase transitions as events from solid to liquid to gas. At all its points of transition, we see the "becoming" of water, its affects, "what it can do" as it joins with other processes: "to flow, to become denser while getting colder yet to expand while freezing, to float as ice, to boil, to rain, to snow, to sleet, to hail, to fog," and so on.

These physical affects are mostly of interest to us here in this chapter as they enter into bioassemblages. Thus we have to see how the water multiplicity is "perplicated" with multiplicities of other natural cycles, involving all the "spheres" that contemporary geographers talk about: the hydrosphere, of course, but also the lithosphere, atmosphere, and biosphere (Wood 2004). However, as Deleuze and Guattari delight in saying after explaining some very complex point, all this talk of "spheres" is "still too simplified" (e.g., Deleuze and Guattari 1987, 56). While the spheres are de jure distinct, they are de facto mixed. Not only are they composed of immensely complex nested sets of coupled cycles at many scales but their intersection zones are also intermixed. For instance, the atmosphere is not a collection of gases but is better thought of as "air," and air has plenty of organisms (spores, microbes), minerals (dust), and water in it. Similarly, the hydrosphere is not just chemically pure H_2O but is "water," which has plenty of organisms, air, and minerals in it. The lithosphere in turn is not just minerals, but its top layer is "soil," which has plenty of air, water, and organisms. Finally, the biosphere's organisms are made of water and minerals and cycle air through them. In Louisiana, I can testify, the air one breathes is an unholy mixture of soil, water, and organisms, the vaporized bayou wafting into your lungs: sometimes you just do not know if you are breathing, drinking, or chewing.

There are other ways in which biological and elemental cycles are linked, as in the way calcium is taken up into animal bones or iron into hemoglobin, but let us just talk about the biological part of the carbon cycle: photosynthesis and respiration. Photosynthesis takes CO_2 and H_2O and uses solar energy to produce carbohydrates and O_2. Respiration takes O_2 and releases the energy used in the chemical bonds of carbohydrates, releasing H_2O and CO_2 back into their cycles. As we will see in more detail later, it is now a truism that bioenergy is just hydrocarbon-mediated solar energy. So you can see the biological link of the water and carbon cycles; you could even say that organic life is just an eddy in the flow of these and many other elements. In sum, organic life uses solar energy to tap into these elemental cycles, to capture and hold some of these flows in other, smaller, and tighter cycles that make up the "organic syntheses" Deleuze (1994, 73) talks about in the discussion of the first synthesis of time in *Difference and Repetition* (see also chapters 8 and 9).

To appreciate the importance of the imbrication of life and the elemental cycles, let us recall the dual definition of *life* in *A Thousand Plateaus*: (1) a set of beings or "organisms," that is, "a particularly complex system of stratification" (Deleuze and Guattari 1987, 336); (2) the creativity of complex systems, "a surplus value of *destratification . . .* an aggregate of consistency that disrupts orders, forms, and substances" (336; emphasis original). Deleuze and Guattari here distinguish organismic life that relies on autopoiesis (the process of maintaining a membrane–metabolism recursivity that conserves an already established pattern of imbrication of the biological and the elemental; see chapters 7–9) and evolutionary change as creative production of a new pattern of biological and elemental imbrication. Seeing the elements as the inorganic strata and the biosphere as the organic strata, then, life as the surplus value of destratification, or in other terms, life as evolutionary change via creative "involution" (238; see Ansell-Pearson [1999, 165–66] for the connection with the work of Lynn Margulis) comes from a novel imbricating of the cycles, that is to say, evolutionary change comes about when microcycles in the hydrosphere, lithosphere, and atmosphere are coupled in a new way with organismic development in the ever-changing biosphere (see Gilbert [2001] and West-Eberhard [2003] on what we will call "eco-devo-evo" in chapter 10).

In stressing creative life as destratification qua new patterns of bioelemental imbrication, we heed the admonition of *A Thousand Plateaus* to avoid any "ridiculous cosmic evolutionism" that would put life on top of, rather than imbricated with, the elemental cycles (Deleuze and Guattari 1987, 49). Let us also note here in discussing the imbrication of life and the elemental cycles that the Gaia hypothesis need not take the extreme position often attributed to James Lovelock, in which Gaia is an organism. Rather, Lynn Margulis's "ecosystem" perspective coupled with her "serial endosymbiosis theory" stance on evolution is much more defensible—and much more Deleuzian, for that matter, for it relies on a notion of Gaia as a multiplicity or set of linked rates of change of differentially defined factors (the different "spheres") punctuated by singularities (the "events" of profound geoevolutionary change, e.g., the oxygen explosion and the rise of the aerobic bacteria as altering both the chemical composition of the atmosphere and the makeup of the biosphere) (Margulis 1998, 119).

More detail on hydrobiology will help us understand Deleuze and Guattari's call for geohistory linked to creative, evolutionary life. In addition to the destratification involved in the imbrication of the biosphere and the elemental cycles, we see an interesting illustration of the interplay of re- and deterritorialization in evolution. Whereas stratification is the process of producing layers of homogeneous bodies, territorialization is the production of links of habits of bodies with features of their environment. In *A Thousand Plateaus*, Deleuze and Guattari (1987, 55) give the following as an example of a line of flight: "When the seas dried, the primitive Fish left its associated milieu to explore land . . . now carrying water only on the inside, in the amniotic membranes protecting the embryo." This is an exemplification of the principle that "an organism that is deterritorialized in relation to the exterior necessarily reterritorializes on its interior milieus" (54). This deterritorializing, developmentally led evolution, this instance of eco-devo-evo, is creative life caught on the fly. Instead of staying in its territorialized way of life, its habits linked to the milieus with which it interacts, the fish folding water inside itself and moving onto the land is forming new habits and, with them, a new, internal milieu. So life is not just "organismic" in the sense of conserving its autopoietic organization; it can be not simply the maintenance of patterns but also a change of patterns. Following one of the "theorems of deterritorialization" of *A Thousand Plateaus*, it is always on the most deterritorializing factor that reterritorialization occurs, even when, in the case of the absolute deterritorialization of nomadism, that reterritorialization is on deterritorialization itself (381, 384). In fact, in the following hydropolitics section, we will see the way the portable water skin allows for this recursive relation of nomadism, how it enables the nomad to be at home while breaking habits and forming a new milieu. But before that, we have more to see in pursuing the hydro-bio-ontology theme.

Sticking with the notion of deterritorialization, in the concept of *Hypersea*, we see that the environment of life on land is the deterritorialized sea (McMenamin and McMenamin 1994; see also Wood 2004, 120–22; Margulis 1998, 109). In a memorable image, McMenamin and McMenamin (1994) tell us that organisms are "lakes" of Hypersea, separated by membranes and connected by ingestion, sex, parasitism, and other forms of communication: "The appearance of complex life on land was a major event in which a kind of mutant sea invaded the land surface. . . . The land biota represents not simply *life* from the sea, but a variation of the sea itself" (25). What is different about multicellular Hypersea organisms is that they have to create systems in which previously independent organisms stick closely together in tightly bound systems enclosed by a membrane that replicates in an enclosed space the organic functions that are distributed in the sea. "Organisms, which are all primarily water, can interact at arm's length, so to speak, only in water. On land, direct physical connections become essential. Overall, terrestrial organisms had to build for themselves structures and components that could perform the environmental services that marine organisms can take for granted" (4). Because of this self-contained structure, "bodies of macroscopic terrestrial plants and animals are the setting for extremely active, if miniaturized, ecological interactions. . . . These interactions constitute Hypersea" (13). The most elementary of these ecosystems, of course, is the eukaryotic cell, as we see in the serial endosymbiosis theory of Lynn Margulis: the mitochondria were originally oxygen-using bacteria that, under the pressure of the "oxygen holocaust," came to live together with other cell elements, providing energy to the emergent unity, the nucleated cell (Margulis 1998, 42). A question, then, for a certain type of political physiology: are the mitochondria slaves or partners?

Hydropolitics

We could go on in this way exploring physiology as politics, but let us shift to think politics as physiology: the body politic as a body, a system of material flows. The State as apparatus of capture sits on top of organic apparatuses of capture so that the State is always seeking control of flows, including flows of water (Deleuze and Guattari 1987, 424–73). To recall the imbrication of the inorganic, organic, and social strata, remember that because water is such a great solvent, it dissolves rock and picks up minerals. Thus, unfortunately for land plants and animals, most of the water on earth—that in the oceans—is too salty. Although we are "hypersea," we are much more dilute than seawater, so we need freshwater; we will supply the minerals in carefully controlled doses by taking in food and drink. How humans have directed freshwater from where there is a lot of it (rivers and aquifers) to where we can use it for drinking or feeding to plants and animals (agriculture) (the process of irrigation) is an important theme we can call *hydropolitics*.

To return to Wittfogel's highlighting of the State–water relation, we should recognize that despite his failings, he does point us to an important truth: aridity is a key factor in the connection of stratified societies and irrigation. Studies on the American West show how the large-scale state and federal investment in irrigation could only produce stratified societies in arid conditions, where control of water grants a key power position (Worster 1985; Reiser 1993). Although Deleuze and Guattari (1987, 363) affirm that "there is no going back on Wittfogel's theses on the importance of large-scale waterworks for an empire," they do acknowledge that some parts of Wittfogel's work have been "refuted" (19). Although they do not enter the details of this refutation (e.g., McAdams 1966, 66–74), when we do, we find that they, curiously enough, affirm some of Deleuze and Guattari's central theses on the State.

Let us take the example of Egypt, as laid out in Butzer (1976). Ancient Egyptian irrigation was basin irrigation rather than canal irrigation. In basin irrigation, earth banks run parallel and perpendicular to the river, creating basins. Sluices would direct floodwater into a basin, where it would sit for a month, until the soil was saturated. Then the water would be drained to the next basin, and the soil in the first basin would be ready for planting. This system sustained Egypt's remarkable continuity (the only ancient irrigated society to have a continuous existence). Once-a-year planting did not deplete the soil, which was replenished by the next year's flood. Nor did basin irrigation result in salination, as the water table during the dry season was well below the root level, so that floodwaters would push accumulated salts down into the water table, below the root level. Butzer thus shows how basin irrigation using the Nile floods arose as a decentralized, locally controlled system and was later overcoded by the apparatus of capture of the State. Butzer writes,

> All of the information that can be brought to bear on Dynastic land use in Egypt shows a simple pattern of winter agriculture,

largely confined to the flood basins, with their crude but effective system of annual flood irrigation. Despite the *symbolic association* of the pharaoh with this inundation [italics added; read "overcoding"], Dynastic irrigation technology was rudimentary and operated on a local rather than national scale. . . . Perhaps the only centralized aspect was the traditional link between tax rates and the potential harvest [State as "apparatus of capture"], as inferred from the height of each Nile flood. . . . No form of centralized canal network was ever achieved in Dynastic times. (50)

In this same vein, we can talk about Stephen Lansing's (2006) work in Bali, which also shows local, decentralized control of canals in the mountains.

With Butzer and Lansing, the contours of the hydropolitical multiplicity begin to come into focus, though to avoid the specter of geographical determinism, we can readily admit that there are multiple pathways (Allen 1997, 53; Gaddis 2004; Davies 2005). The multiplicity behind the morphogenesis of hydropolitical structure includes geological factors such as ground slopes and surface friction; biological factors such as type and strength of local flora and fauna; and hydrological factors such as river currents, channels, and wave strengths. In addition, as we will explore later in the next section and in chapter 3, it also includes social–technical factors such as the speed capacity of warfare assemblages—the phalanx as man–spear–shield assemblage; the chariot as horse–men (driver and fighter)–bow assemblage—and the waterborne assemblages of rower-powered warships and sailing-power merchant ships. Wittfogel's mistake was seeing a single determining factor—control of irrigation—in the morphogenesis of Oriental despotism. This is not to say that it was not important, just that it could not determine the pathways to empire.

With all the warnings against a simplistic geographical determinism in mind, we can still sketch out some of the key dimensions of the multiplicity underlying the development of the river valley empires, building on the insight into interlocking factors in McAdams (1966, 46–47): "trends toward territorial aggrandizement, political unification, and population concentration . . . [are] . . . functionally integrated processes." So we can look for flat lands allowing for irrigation-intensive agriculture and the installation of garrisons in outlying towns that can enforce the exploitation of tribute labor; we can see that the corvée produces roads as well as irrigation and monuments. Once past a certain threshold, we find a positive feedback loop: the bigger the territory under control, the more solar energy is captured in agriculture and the larger the bureaucracy and the army that can be fed with the surplus. These can then enlarge and administer the territory and put more peasants to work producing and funneling surpluses and building roads for more expansion, and so on, up to the limit of information bottlenecks, trade-offs between speed of transport and size of territory, and so on. It is important to avoid an economic determinism that would privilege a food surplus (in our terms, the solar economy) as independent variable and political development as dependent (for critiques of the search for independent variables in social science, see Gaddis [2004] and Bonta and Protevi [2004]). As McAdams (1966, 47) puts it quite clearly, historical accounts of the development of the Mesopotamian empires require us to look for interlocking solar-techno-political processes:

> Extensions of territorial control, new forms of political superordination, and a multiplicity of technological advances all may have had as much effect on the size of the [agricultural solar energy capture] surplus as improvements in immediate agricultural "efficiency," while the deployment of the surplus, however it was formed, obviously had important reciprocal effects on these other factors.

McAdams sums it up quite nicely: the development of centralized empires in Mesopotamia relies on "an interdependent network of cause and effect" (47).

Butzer (1976), in fact, shows that in Egypt, the key factor for Pharaonic absolutism was the ecological embeddedness of *nomes* or basic territorial–political social units. "These primeval nomes appear to have provided the necessary political infrastructure for the military ventures that over several generations of strife led to the unification of Egypt. In this sense Pharaonic civilization remains inconceivable without its ecological determinants, but not in the linear causality model [sc. Wittfogel] of stress → irrigation → managerial bureaucracy → despotic control" (111). In other words, Butzer does not deny that Egypt was united under a despot nor that its political structure was ecologically embedded. He just denies that irrigation control was the sole determinant of that imperial-scale despotism. Here we see an important question of scale: for Butzer, the nome or local unit qualifies as a State—perhaps even a hydraulic state—but not as an empire. In fact, it has recently been argued that centralized national state control of Egyptian irrigation—based on a change from basin irrigation to a centralized canal grid system—is a nineteenth-century phenomenon that was represented as a return to the supposedly centralized irrigation control of the Pharaohs. So the argument would be that with regard to Egypt, at least, Wittfogel mistook modern propaganda for ancient reality (Kalin 2006). Butzer (and, by extension, Lansing) thus contradicts Wittfogel, who stresses the State as the origin of large-scale waterworks, and confirms Deleuze and Guattari's theses that the imperial State overcodes local arrangements.

We should recognize in conclusion to this section, however, that the State has no monopoly on hydropolitics, for there is a "hydraulic model of nomad science and the war machine . . . [that] consists in being distributed by turbulence across a smooth space" (Deleuze and Guattari 1987, 363). We need to add to Deleuze and Guattari's discussion of the nomad to bring ontology and politics more closely together in the study of hydropolitics. For the portable water container, the animal skin, is as fully a part of the nomad assemblage as the more famous stirrup, and the machinic phylum had to encompass this technological supplement to Hypersea to allow the nomad smoothing and occupation of the arid steppes.

Bio-solar-political Physiology

In the preceding, we talked about the earth and about water and how they link to biology and politics; here we can talk about the sun and about war in the way they are linked in supra- and subpersonal registers in political physiology.

About the sun, one need not credit all of Bataille's more melodramatic statements to recognize that looking at circuits of solar energy as the basis of life is a staple of introductory biology textbooks (e.g., Mader 2009). And certainly, looking at the ways societies waste excess in wars or monuments via Bataille's notion of *dépense* or radical expenditure is the key to Deleuze and Guattari's grappling with the Marxist questions about capitalist crises of overproduction and the "realization of surplus value" (Deleuze and Guattari 1984, 4n, citing Bataille 1991). Scarcity is not natural, but is produced, via *dépense*, Deleuze and Guattari claim, to instill fear of starvation–isolation (friendship can be made scarce by shunning—and friendship is as physiologically necessary as food, even if you die of loneliness more slowly than you starve) that will reinforce social structures that code flows (28). Concerning war and political physiology, finally, I define that term as the study of the construction of bodies politic, that is, the interlocking of emergent processes that link the patterns, thresholds, and triggers of affective and cognitive responses of somatic bodies to the patterns, thresholds, and triggers of actions of social bodies. Political physiology has a wider application than consciousness studies because political institutions interlock with individual physiology in emotional responses to commands, symbols, slogans, and images; such responses at least strongly condition actions, through unconscious emotional valuing, but sometimes—even if rarely—provoke behavior that completely eludes conscious control, as in depths of panics and rages (Protevi 2009). Political physiology also asks us to look at the act of killing and its relation to political sovereignty (chapter 2). The traditional definition of sovereignty is that it is vested in the political body that holds the monopoly on the legitimate use of force within a clearly defined geographical territory. Thus, at the limit, a political body must be able to control the triggering of killing behavior in the bodies of its "forces of order" (army and police).

In principle, we could look to the articulation of political physiology and biosolar politics in contemporary life, in the two Iraq wars and their bizarre American offspring, the Hummer, where global petroleum wars meet the anxious individual driving an armored car around suburbia. But here I will focus on the ancient Eastern Mediterranean. We can begin with some geopolitical basics. We discussed earlier some factors in the genesis of ancient Mesopotamian empires involving irrigation-intensive agriculture in flat river valleys. Again, keeping warnings against geographical determinism in mind, we can nonetheless point to the traditional idea that *poleis* benefited from mountainous terrain to maintain independence, each mountain range enclosing a farming region supporting small farmers who were able, by forming a phalanx of armored hoplites, to overcome aristocratic dominance and demand *isonomia* or equality before the law (Sacks 1995, 101, 190). I am, of course, here treating very complex matters with a very broad brush; nonetheless, the hoplite–equality connection is, for all its simplicity, well attested to in even the most technical discussion aiming to eliminate the too-crude notion of a "hoplite revolution" and to pin down the role of hoplites in supporting tyrants who break aristocratic dominance (Raaflaub, Ober, and Wallace 2007, 34–36, 121, 133–36; Ste. Croix 1981, 260, 282; 2004, 126–27).

It may seem odd at first, but from a solar-politics perspective, we can claim olive oil as a key factor in the genesis of Athenian democracy. Olive oil is a storage form of solar energy burned for light in lamps and burned for energy in human bodies. One of the tipping points in the democratization process in Athens occurred when Solon forbids debt slavery and debt bondage (Raaflaub, Ober, and Wallace 2007, 59; Ste. Croix 1981, 137, 282) as well as all agricultural exports except that of olive oil. This last provision stabilized the middle class of small farmers, who were threatened by aristocratic dominance (Milne 1945; Molina 1998). These farmers produced olive oil as a cash crop (a small part of their total production, to be sure: it was the large farmers who dominated the oil market; nonetheless, it was a crucial money source). This stabilization of a mass olive oil export market created demand for work by urban artisans who produced jars for olive oil and manufactured goods for export (also arms for hoplites to forestall aristocratic reconquest). A growing urban population needed grain importation, however, and protecting the import routes needed a naval force. In turn, what we can call the *military egalitarianism* thesis retained its force and claimed that a dependence on a naval force pushed the regime toward urban democracy, that is, toward expanding the political base beyond that of the hoplites, for rowers were drawn from ranks of urban masses unable to afford hoplite gear (Raaflaub, Ober, and Wallace 2007, 119–36; Gabrielsen 2001).

Now democratic rowing in the Athenian navy (leaving aside the question of seaborne marine troops) was relatively low intensity, at least compared to the hand-to-hand fighting depicted in Homer and the phalanx clashes of the classical age. (Actually, to respect the Deleuzian emphasis on assemblages, we should note that "hand-to-hand" is a misnomer, for shield and sword–spear is itself quite a bit less intense than just one-on-one with hands.) Thus, for rowers, there was less necessity for the high-intensity training needed for noble single combat. To produce such a warrior body, you need to traumatize it by lots of intensive hunting and fighting as boys; here we can think of Odysseus's scar from his adolescent rite of passage, the boar hunt (*Odyssey* 19.390–95). Phalanx training was intermediate between aristocratic single combat and naval rowing; it was less intense than single combat because of teamwork, that is, emergence. In the phalanx, you stood by your comrades rather than surging ahead. Recall Aristotle's definition of *courage* as the mean between rashness and cowardice; in comparative terms, rashness for the hoplite, the phalanx soldier, was standard behavior for the Homeric warrior, whereas phalanx courage—staying with your comrades—would be mediocrity if not cowardice for the warrior (*Nicomachean Ethics* 2.8.1108b20–27).[2]

The discrepancy between Aristotelian and Homeric courage is an excellent example of the Deleuzian distaste for essentialism: you will never come up with a set of necessary and sufficient conditions to define *courage,* so it is much better to investigate the morphogenesis of warrior and soldierly bodies and see if there are any common structures to those production processes. The question we want to ask from this perspective is, how are the warrior and the soldier different actualizations of the virtual multiplicity linking political physiology and hydro-solar-politics? And this standing together is the key to the *eros,* the ecstatic union with a social body, of the phalanx. As we will see in a moment, William McNeill's (1995, 117) *Keeping Together in Time* allows us to account for this human bonding in terms of collective resonating movement provoking the entrainment of asubjective physiological processes supporting emotional attachment.

But before we go below the subjective level to entrainment and political physiology, we should note its complement in the suprasubjective materialist explanation of Athenian foreign policy by the noted Marxist historian G. E. M. de Ste. Croix. In *Origins of the Peloponnesian War* (Ste. Croix 1972) and again in *Class Struggle in the Ancient*

Greek World (Ste. Croix 1981), Ste. Croix points out that the geopolitical key to the transition from Athenian democracy at home to the Athenian Empire after the Persian Wars is the threshold of human energy production from grain ingestion. Ste. Croix (1981, 293) uses this to undercut ideological explanations of Athenian foreign policy: "I have . . . explained why Athens was driven by her unique situation, as an importer of corn on an altogether exceptional scale, toward a policy of 'naval imperialism,' in order to secure her supply routes." The singularities in the Athenian actualization of the geo-hydro-bio-political multiplicity are what gets us out of ideological condemnations of a supposed Athenian "lust for power." As Ste. Croix (1972, 47–49) points out, rower-powered warships had a much shorter range than sail-driven merchant ships, which are able to capture solar energy in the form of wind power—itself generated from a multiplicity of temperature differentials of land mass–sea–water currents producing wind currents (see also Gomme 1933). So the Athenian democrats needed a network of friendly regimes whose ports could provide food and rest for the rowers of their triremes, that is, to use our terminology, to replenish the biological solar energy conversion units of the triremes qua "man-driven torpedo[es]" (Gabrielsen 2001, 73). Bringing the geo-solar-hydrological dimensions of the multiplicity together with biotechnical and more traditionally sociopolitical dimensions, a recent scholarly article puts it this way: "the concept of *thalassokratia* [sea power] implies intense naval activity, primarily in order to defend existing bases and to acquire new ones, and intense naval activity, in its turn, requires command over enormous material and financial resources" (Gabrielsen 2001, 74).

Adopting this viewpoint allows us to understand the suprasubjective and anti-ideological materialism of a key passage from Ste. Croix (1981, 293; emphasis added): "Athens' whole way of life was involved; and what is so often denounced, as if it were sheer greed and a lust for domination on her part, by modern scholars whose antipathy to Athens is sharpened by promotion of democratic regimes in states under her control or influence, *was in reality an almost inevitable consequence of that way of life.*" But we should not be content with only going above the subjective; we should go below it as well. McNeill's reading of the political consequences of entrainment-provoked military solidarity takes us below the subject, complementing Ste. Croix's suprasubjective geomaterialism. McNeill (1995, 117) writes, "The Athenian fleet developed muscular bonding among a larger proportion of the total population than ever fought in Sparta's phalanx." Furthermore, "feelings aroused by moving together in unison undergirded the ideals of freedom and equality under the law. . . . The muscular basis of such sentiments also explains why the rights of free and equal citizens were limited to the militarily active segment of the population" (118).

Let me conclude with this speculation that brings together Gabrielsen's adjunct-subjective biotechnical assemblage (the trireme as "man-driven torpedo"), Ste. Croix's suprasubjective geopolitics, and McNeill's subsubjective entrainment-provoked emotional solidarity. Let us see these as the dimensions of a hydro-solar-bio-techno-political multiplicity whose democratic naval actualization in Athens may help to explain why Plato in the *Laws* (Hamilton and Cairns 1961) put the ideal city away from the sea and why, in the same dialogue, hoplite victory in the land battle of Marathon is praised over democratic rowers' victory in the sea battle of Salamis. "We . . . insist that the deliverance of Hellas was begun by one engagement on land, . . . Marathon, and completed . . . at Plataea. Moreover, these victories made better men of the Hellenes, whereas [Salamis] did not." Plato concludes in words that at least point to our concern with geo-hydro-solar-bio-technical-political multiplicity: "In our investigations into topography [*chôras phusin*] and legislation [*nomôn taxin*] [we focus on] the moral worth of a social system [*politeias aretên* or "political excellence–virtue"]" (707c–707d, citing the translation in Hamilton and Cairns 1961).

≈ 2 ≈

The Act of Killing in Contemporary Warfare

This chapter continues the investigation, in the context of warfare, of the relation of the social and the somatic in the production of bodies politic. That is, I look to the spiraling diachronic mutual presupposition that produces the transgenerational fit of subjectification practices and affective cognitive makeup. Here I focus on the interplay between the subsubjective neurophysiological and the adjunct-subjective technical more than in chapter 1, in which I looked more to the interplay of the suprasubjective geopolitical and the adjunct-subjective technical. Chapter 3, the conclusion to part I, achieves more of a balance of going "above, below, and alongside the subject."

Although it is a similar exploration of the internally differentiated multiplicity of the geo-bio-techno-political as it is individuated in ontogenetically effective subjectification practices producing differenciated (type-similar) bodies politic (which are in turn the field of individuation for behavioral acts), the chapter's emphasis on contemporary practices forms a chronological counterpoint to the emphasis on ancient warfare in chapters 1 and 3. Note, however, that even if the training that produces a militarized, combat-ready body politic is different in different regimes, the fact that there must be training shows that the preparation for the intensity of combat must be seen against a fairly stable and widespread evolutionary backdrop of inhibitions on close-range killing. Of course, in discussing human evolution, we must always keep in mind the need to look for variation in populations in both genetic and epigenetic factors—"difference and development," as I put it in chapters 4–6—as even simple in-group violence-inhibiting empathy is dependent on some minimally nonabusive upbringing.

The Act of Killing

Killing in combat is less easy than it might seem to those outside the military, for whom the logic of "kill or be killed" would predict high rates of deadly interaction. While close-range killing can be done by a very small percentage of soldiers in cold blood (with full conscious awareness of a subject), Grossman (1996) argues for a deep-seated inhibition against one-on-one, face-to-face, cold-blooded killing on the part of some 98 percent of soldiers, a figure that correlates well with the estimated 2 percent of the population who count as low-affect or "stimulus-hungry" sociopaths (Niehoff 1999; Pierson 1999).

Grossman's qualifications with regard to distance, teamwork, command, and mechanical intermediaries are dimensions of the multiplicity for the act of killing. Certain values in the relation of rates of change of these variables will, in suitably ontogenetically trained bodies politic, serve as triggers that increase the ability to engage in deadly combat. In looking at different training regimes, researchers were at first shocked and dismayed to find that traditional military drill (target shooting at bull's-eyes) produced only a 15 to 20 percent *firing* rate among American infantry troops in World War II, excluding machine gunners (Grossman 1996, 3–4, citing Marshall 1978; see also Collins 2008). Now a firing rate doesn't indicate willingness to kill, as Grossman explains. The usual "fight or flight" dichotomy is falsely drawn from *inter*-species conflicts; *intra*-species conflicts are also marked by display and submission, which, along with flight, are much more likely to occur before fight (especially fight to the death). While it is true that in some territorial species, such as lions, a newly victorious alpha male will kill the offspring of his defeated adversary, the intraspecies inhibition we invoke concerns animals of the same generation in one-on-one combat; chimpanzee wars and murders always involve ambushes in which at least two, but often seven or eight, chimpanzees attack a single, isolated victim (de Waal 1997, 38). Given these factors, Grossman concludes that much of the World War II firing rate was display rather than fight (Grossman 1996, 5–6).

I propose two factors to account for the wide distribution of the inhibition on killing among humans, each of which depends on what is at least a proto-empathetic identification.[1] We need not decide here on the mechanism of that empathy, for which there are two major explanations in the current literature. First, we find simulation theories. The by-now classic scholarship here does not rely on action-oriented mirror neurons (as Vittorio Gallese [2001] thought in his "shared manifold" article) but on what Gallese, Keysers, and Rizzolatti (2004) call *viscero-motor centers*; here they refer to the findings of Singer et al. (2004), in which "empathy for pain" is correlated with increased activity of the anterior insula and the anterior cingulate cortex. Second, we find phenomenological accounts. Some supplement simulation theory with an account of an embodied intentionality, as in Thompson (2001); others, however, will find the simulation theory approach still too representational and appeal to a field of directly felt corporeal expressivity or "primary embodied intersubjectivity" grounding our "pragmatic interaction" with others (Gallagher 2005, 223). The phenomenological approach finds support in the widespread recognition of the humanity of the opponent through the sight of the face. Face recognition is one of the earliest infant capacities (Hendriks-Jansen 1996, 252–77; see also Stern 1985; Gallagher 2005), and many battlefield accounts show how the face of the enemy has profound inhibitory effects; the blindfold on the victim of a firing squad enables the shooters by breaking eye contact between victim and executioners (Grossman 1996, 225).

Either approach seems superior in accounting for this inhibition on killing to *theory theory* (agreeing here with the emphasis on affect found in Maibom [2007]). Rooting the sort of intense identification we find in cases of anticipation or recollection of close-range killing—consider the trembling limbs, the intense nausea, the bouts of vomiting that we find here—in a cognitive inference via the observation or anticipation of outward behavior (in this case, writhing in agony and clutching at spilled guts), so that we *attribute* the emotional state of agony to the *mind* of another person, seems rather thin soup—thin soup akin to the folk cognitivism of popular media accounts that describe waterboarding as the production of the belief in the mind of the victim that he or she is drowning.[2] We should rather describe waterboarding as triggering an evolutionarily preserved panic module that acts by means of a traumatizing biochemical cascade. We here see the link of internalist or cognitivist approaches and a certain "neurocentrism": it's only by bracketing the endocrine system in favor of an exclusive focus on the central nervous system (and there, focusing on electrical activity somehow abstracted from its biochemical milieu of neurotransmitters and hormones) that one could think of "beliefs" here.

Whether it is a simulation or an embodied intersubjectivity, there is a fundamental linkage of affect, body image, and bodily integrity in the experience of proto-empathic identification. Soldiers' testimonies are clear that seeing someone else's blood and guts spill out of them is powerfully felt by many soldiers (Kirkland 1995; Kilner 2000). These phenomena invite a concept of "political physiology"—the linkage of social and somatic—that can shed light on pornography and thanatography (the bodily reaction to images of violence) and their intersection in the Abu Ghraib scandal as well as in the current wave of "torture porn" films (the Saw and Hostel series, in particular). Susan Hurley (2004) broached this topic in her work on media violence.

Whatever the mechanism, these proto-empathetic abilities are two possible factors underlying the widespread inhibition on cold-blooded close-range killing: (1) sensing what the intensity of the fight to the kill would be like—an attack beyond the threshold mutually recognized as that indicating display might provoke a deeply panicked self-defense on the part of the opponent rather than the desired submission[3]—and (2) the need to avoid the intensity of revulsion afterward: living with having been a killer would be too much; the "memories of the future" (Casey 2000, 62–63) are in this case intolerable for the subjective present (see also Damasio [1994] and [1999] for accounts of the role of "somatic markers" in entertaining as-if scenarios of future action).

Rage, Ancient and Modern

As we have seen, the vast majority of soldiers cannot kill in cold blood and need to kill in a desubjectified state, for example, in reflexes, rages, and panics. But who does the killing when reflexes, rages, and panics are activated?

In recent works, Dan Zahavi (2005) and Shaun Gallagher (2005) distinguish agency and ownership of bodily actions. *Ownership* is the sense that my body is doing the action, whereas *agency* is the sense that I am in control of the action, that the action is willed. Both are aspects of subjectivity, though they may well be a matter of prereflective self-awareness rather than full-fledged objectifying self-consciousness. But alongside subjectivity, we need also notice emergent assemblages that skip subjectivity and directly conjoin larger groups and the somatic. To follow this line of thought, let us accept that in addition to nonsubjective body control by reflexes, we can treat basic emotions as modular "affect programs" (Griffiths 1997) that run the body's hardware in the absence of conscious control.[4] As with reflexes, ownership and agency are only retrospectively felt, at least in severe cases of rage, in which the person "wakes up" to see the results of the destruction committed while she was in the grips of the rage. In this way, we see two elements we need to take into account besides the notion of subjective agency: (1) that there is another sense of "agent" as nonsubjective controller of bodily action, either reflex or basic emotion, and (2) that in some cases, the military unit and nonsubjective reflexes and basic emotions are intertwined in such a way as to bypass the soldiers' subjectivity qua controlled intentional action. In these cases, the practical agent of the act of killing is not the individual person or subject but the emergent assemblage of military unit and nonsubjective reflex or equally non-subjective affect program.

A little more detail on the notion of a *rage agent* might be helpful at this point. Extreme cases of rage produce a modular agent or *affect program* that replaces the subject. Affect programs are emotional responses that are "complex, coordinated, and automated . . . unfold[ing] in this coordinated fashion without the need for conscious direction" (Griffiths 1997, 77). They are more than reflexes, but they are triggered well before any cortical processing can take place (though later cortical appraisals can dampen or accelerate the affect program). Griffiths makes the case that affect programs should be seen in light of Fodor's notion of modularity, which calls for a module to be "mandatory . . . opaque [we are aware of outputs but not of the processes producing them] . . . and informationally encapsulated [the information in a module cannot access that in other modules]" (93). Perhaps second only to the question of adaptationism for the amount of controversy it has evoked, the use of the concept of modularity in evolutionary psychology is bitterly contested. I feel relatively safe proposing a very widely distributed rage module or rage agent because its adaptive value is widely attested to by its presence in other mammals and because Panksepp (1998, 190) is able to cite studies of direct electrical stimulation of the brain and neurochemical manipulation as identifying homologous rage circuits in humans and other mammalian species. Panksepp proposes as adaptive reasons for rage agents their utility in predator–prey relations, further sharpening the difference between

rage and predator aggression. Whereas a hunting attack is by definition an instance of predatory aggression, rage reactions are a prey
phenomenon, a vigorous reaction when pinned down by a predator. Initially a reflex, Panksepp claims, it developed into a full-fledged
neural phenomenon with its own circuits (190). The evolutionary
inheritance of rage is confirmed by the well-attested fact that infants
can become enraged by having their arms pinned to their sides (189).

With this as background, let us concentrate on the individuation
of rage states in suitably trained military personnel. Recalling the
dimensions of the bio-technical-affective multiplicity underlying the
individuation of acts of killing, we see that without the enablers of
proper levels of distance, machinics, teamwork, command, and dehumanization, most soldiers must leave the state of "cold blood" to kill
one-on-one at close range—they have to dump their subjectivity. They
burst through the threshold of inhibition by supercharging their bodily
intensity. Thus the tried-and-true method for killing in close combat is
the berserker rage, the frenzy of killing anything that enters the "death
zone" immediately in front of the berserker. In the berserker rage, the
self-conscious and controlled subject is overwhelmed by a chemical
flood that triggers an evolutionarily primitive module that functions
as a rage agent, running the body's hardware in its place. The Greeks
called it "possession by Ares" (Shay 1994; Harris 2001). It is important
to understand that such rage is itself traumatic: it sets endorphin
release thresholds so high that only more combat will provide relief,
initiating a cycle of rage, trapping many of those who enter it in the
berserker state and greatly increasing the chance of post-traumatic
stress disorder (PTSD) (van der Kolk and Greenberg 1987; Shay 1994).

A common trigger of the berserker rage is the death of a comrade
(Shay 1994; Kirkland 1995). We can speculate that such rage is triggered by what Damasio would call the flashing somatic marker of
future pain (separation from and mourning for the comrade) coupled
with the memory of pleasure tagged to the person of the comrade.[5]
The wrenching shift between the pleasant memories and the painful
future triggers rage, a notion that dovetails with Panksepp (1998),
in which frustration, as the curtailment to the free use of "seeking"
and "play" systems, triggers rage. Another trigger, at which we have
already hinted, is direct and immediate threat to life, the panicked

self-defense reaction that display and submission seeks to avoid. There are, of course, many other rage triggers in other walks of life we can't discuss here, among them abandonment, as when domestic violence escalates from beating to killing, as often happens only after separation. The military problem of the berserker rage is how to turn it on and off on command (and only on command): this is the problem of the conversion of the warrior (whose triggers include insults to honor) into the soldier who kills only on command.

The military problem is that rage or panic agents have no emergency brakes. For example, the ancient Norse berserkers were very effective killers but could not stop killing at will; their berserker state was only turned off once all members of the opposition were dead (Speidel 2002). We can note that modern soldiers are not *trained* to utilize rage states; the goal of modern military training is not to replicate the berserkers of ancient times. As we will shortly see, for most modern soldiers, the attack direction is articulated to neither the panic nor the rage agent but, in free-fire zones, to the conditioned response of the sight of a human silhouette, or in urban warfare situations, to key traits in the appropriate context. Occasionally, however, rage and panic agents can supersede controlled, that is, circumscribed and predictable, reflex killing, if the situation involves the death of a comrade (Shay 1994).

Even when a sense of agency is absent during the rage-induced or reflex-controlled act of killing, however, a sense of moral responsibility can be produced by a retrospective identification of action and ownership, a retrospective production of the moral sense of agency, even when the practical agent at the time of action was a nonsubjective rage or reflex: "Oh, my God, look what I've done!" In support of this claim, let us turn to Lifton (1973), who has produced a noteworthy study of Vietnam veterans of the My Lai massacre, in which the psychological trauma of such guilt-producing retrospective identification plays a central role. After discussing the ways in which many aspects of the American war in Vietnam set up an "atrocity-producing situation" (41), he provides a brief description of the "psychology of slaughter," in which rage and racialized dehumanization of the enemy play a major role (42–43). Of particular interest to us is his description of individual soldiers' experiences of guilt after the rage-fueled group performed the slaughter at My Lai (56–57; on the toxic combination of killer and survivor guilt, see 107). Even though we could argue that the practical agent of the massacre was the assemblage of the unit and the distributed nonsubjective rage agents, these soldiers assumed moral responsibility; that is, they identified themselves as individual moral agents with the distributed and emergent practical agent of the massacre. In what follows, we discuss two contemporary modes of military training in relation to the phenomenon of retrospective identification of ownership and agency resulting in guilt.

Two Modes of Contemporary Military Training

Reflex Training

Contemporary military training cuts subjectivity out of the loop so that most soldiers' bodies are able to *temporarily* withstand the stress of the act of killing. The first aspect is affective: soldiers are acculturated to dehumanize the enemy by a series of racial slurs. This acculturation is especially powerful when accomplished through rhythmic chanting while running, for such entrainment weakens personal identity to produce a group subject (McNeill 1995; Burke 2004). At the same time as the group subject is constituted, the act of killing is rhetorically sterilized by euphemisms:

> Most soldiers do not "kill," instead the enemy was knocked over, wasted, greased, taken out, and mopped up. The enemy is hosed, zapped, probed, and fired on. The enemy's humanity is denied, and he becomes a strange beast called a Kraut, Jap, Reb, Yank, dink, slant, or slope. (Grossman 1996, 93)

Desensitization is merely an enabling factor for the role of classical and operant conditioning in modern training. Such training enables most soldiers to kill reflexively. In doing so, they bypass the widespread inhibition on killing we noted earlier.

The major problem of modern military training that reconfigures reflex action lies in going beyond what the restored subjectivity of many soldiers can withstand. The shoot-on-sight or free-fire-zone protocol begins in Vietnam with the application of human silhouettes rather than concentric targets in basic training; this new training produced a significant rise in kill-to-fire ratios (Grossman 1996, 181). In effect, such pattern recognition training increases the distribution of a "hunter agent" in the population of soldiers so that the sight of human-shaped targets triggers a shoot reflex. The problem, here, however, is that the increased distribution of hunting agents is incompatible with the widespread proto-empathetic identification we discussed earlier. Unless this proto-empathetic identification is sufficiently desensitized, many soldiers are psychologically traumatized, because in the aftereffects of battle, they see the enemy's corpse—produced by their implanted hunting agents—as human, as someone "who could have been me" (Lifton 1973; Grossman 1996). In combination with the physiological effects of long-term stress (in particular, elevated cortisol levels), such psychological trauma is linked with PTSD (Shay 1994; van der Kolk and Greenberg 1987).

Cyborg Training

Vietnam-era reflex training is good only for free-fire zones. With urban warfare, more sophisticated cognition is necessary: the shoot–no shoot instant decision. With the advent of digital and video simulator training for urban warfare, we see true cyborg killing.

Military training has very often involved simulated combat conditions—training dummies—to develop motor skills. While it succeeds in this, the transfer to real combat often falters because of affective limitations. Traditional simulation training puts soldiers in an everyday world of three-dimensional objects; however, the difference between the dummy and a real person is clear so that killing the dummy does not desensitize proto-empathetic identification. Digital and video simulation (live action figures with a computer-generated image backdrop) develops individual motor skills, but we can speculate that it also increases the desensitization effect of training. Because images are so lifelike, they activate the proto-empathetic identification present in most. Repetition of the training attempts to produce the desired desensitization. In other words, simulation-trained contemporary soldiers have already *virtually* experienced killing before *actually* having to kill (Macedonia 2002; McCarter 2005). But they haven't experienced the transition from the simulated environment to real life: we speculate that even though simulations can desensitize to some extent, they cannot override or completely extinguish the proto-empathetic identification capacity in a good number of soldiers. (We are dealing with very complex matters here regarding PTSD in the current Iraq campaign [Hoge et al. 2004]. Anecdotal evidence relayed to the author in personal communication by Lieutenant Colonel Pete Kilner of West Point suggests that officers who had talked and thought about the aftereffects of killing had less guilt than enlisted men and women without such preparation.)

In addition to the affective aspect of heightened desensitization, simulation training constitutes a new cognitive group subject. The instant decision of shoot–no shoot is solicited by the presence or absence of key traits in the gestalt of the situation. Such instant decisions are more than reflexes but operate at the very edge of the conscious awareness of the soldiers and involve complex subpersonal processes of threat perception (Correll, Urland, and Ito 2006). In addition to this attenuation of individual agency, cutting-edge communication technology now allows soldiers to network together in real time. With this networking, we see an extended–distributed cognition culminating in "topsight" for a commander who often does not command in the sense of micromanaging but who observes and intervenes at critical points (Arquilla and Rondfeldt 2000, 22). In other words, contemporary team-building applications through real-time networking are a cybernetic application of video games that goes above the level of the subject (Fletcher 1999). In affective entrainment, instant decision making, and cognitive topsight, the soldiers produced by rhythmic chanting and intensive simulation training are nodes within a cybernetic organism, the fighting group, which maintains its functional integrity and tactical effectiveness by real-time communication technology. It is the emergent group with the distributed decisions of the soldiers that is the cyborg here, operating at the thresholds of the individual subjectivities of the soldiers.

What happens to these soldiers once they return home and are no longer part of the larger cybernetic organism that was constituted by their very bodies? What happens to soldiers when they are separated from the group subject, the true practical agent of the act of cyborg killing? We should remark on the tenacity of retrospective guilt produced through the "my God, what have I done?" effect. Even when the practical agent of the act of killing is the assemblage of emergent military unit and distributed nonsubjective reflexes, rage agents, or awareness-threshold decisions, we can see a "centripetal power" to subject constitution, drawing to itself responsibility for acts it never committed in isolation. Thus it seems many soldiers paradoxically just cannot help taking responsibility. In other words, to heighten the paradox, they are irresponsible in taking responsibility, in taking on themselves moral agency, when practical agency lies elsewhere. Questions for future research concern the genealogy of this powerful motivation for subject construction and the assumption of moral responsibility.[6]

⇒ 3 ⇐

Music and Ancient Warfare

This is the final chapter of part I, our final look at the supra-, adjunct-, and subsubjective in the production of combat-ready military groups and individuals. As with chapter 1, the focus will be on ancient warfare, as opposed to the contemporary practices we examined in chapter 2. Conversely, as in chapter 2, I will target the subsubjective neurophysiological register more so than I did in chapter 1, which looked mostly at the interlocking of the suprasubjective geopolitical register of water administration and food importation with the adjunct-subjective technical register of the phalanx and the trireme. I will also highlight the notion of affect as it varies with changes in the assemblages that actualize the geo-bio-techno-political multiplicity. As Deleuze and Guattari (1987, 400) put it in *A Thousand Plateaus*, "passions are effectuations of desire that differ according to the assemblage."

Our guiding thread in this chapter will be the following passage, in which Deleuze and Guattari (1980, 497–98; 1987, 399–400; translation modified) describe the imbrication of the social and the somatic, the technical and the affective, in an assemblage:

> Assemblages [*agencements*] are passional, they are compositions of desire. Desire has nothing to do with a natural or spontaneous determination; there is no desire but assembling, assembled, machined desire [*il n'y a de désir qu'agencant, agencé, machiné*]. The rationality, the efficiency, of an assemblage does not exist without the passions the assemblage brings into play, without the desires that constitute it as much as it constitutes them. Detienne has shown that the Greek phalanx was inseparable from a whole reversal of values, and from a passional mutation that drastically changed the relations between desire and the war machine. It is a case of a man dismounting from the horse, and of the man–animal relation being replaced by a relation between

men in an infantry assemblage that paves the way for the advent of the peasant-soldier, the citizen-soldier: the entire Eros of war changes, a group homosexual Eros tends to replace the zoosexual Eros of the horseman.

We are going to use this passage as a jumping-off point to examine some of the geo-bio-techno-affective assemblages at work in ancient Greek and Near Eastern warfare. We will look at the phalanx, but for greater contrast, we will not focus on the "zoosexual" horsemen but on the berserker "runners" and their putative involvement in the 1200 BCE collapse of the Bronze Age kingdoms, of which Mycenae and Troy are the most famous examples.

The chapter will have two parts. In the first part, I explore the ontology, biology, and history of affect. In each case, we put Deleuze and Guattari's work into the context of current research in cognitive science, biology, anthropology, military history, and biocultural musicology. Exploring the *ontology of affect* will enable us to explain the exteriority of affect versus the interiority of emotion or "feeling." Exploring the *biology of affect* will take us to the biophilosophical school known as developmental systems theory (DST) as it intersects new research in neuroscience. From this perspective, the patterns, triggers, and thresholds of affective cognition are produced via transgenerational subjectification practices. The social and the somatic are not synchronic opposites but are linked in a spiraling diachronic interweaving at three temporal scales: the long-term phylogenetic, the mid-term ontogenetic, and the short-term behavioral. Finally, in exploring the *history of affect,* we will discuss the knotty problems of cultural evolution and the anthropology of war.

In the second part, we deploy this theoretical schema to examine the use of music in ancient warfare. In showing the difference between the types of music used to trigger the berserker rage of the "hill runners" versus the type used to evoke the entrainment-induced solidarity of the hoplite phalanx, we will demonstrate the specificity of bodies politic, the interlocking of supra-, adjunct-, and subsubjective registers in the individuations that provide the differenciations (the types of bodies politic and of their acts) of the internally differentiated multiplicity of geo-bio-techno-political warfare.

The Ontology, Biology, and History of Affect

The Ontology of Affect

For Deleuze and Guattari, "affect" comprises the active capacities of a body to act and the passive capacities of a body to be affected or to be acted on. In other words, affect is what a body can do and what it can undergo. The use of this term derives from Deleuze's reading of Spinoza (Deleuze 1988), in which Deleuze carefully distinguishes "affect" (*affectus*), the experience of an increase or decrease in the body's power to act, from "affection" (*affectio*), the composition or mixture of bodies. (More precisely, affection is the change produced in the affected body by the action of the affecting body in an encounter.) *Affectus*, or what we could call *experiential affect*, is not representational, Deleuze remarks, "since it is experienced in a living duration that involves the difference between two states." Indeed, we should say that as such an experience of difference, *affectus* is "purely transitive" (Deleuze 1988, 49).

In the main discussion of affect in *A Thousand Plateaus*, Deleuze and Guattari do not maintain the Spinozist term *affection*, but they do distinguish the relations of the extensive parts of a body, which they call "longitude," from the intensities or bodily states that augment or diminish the body's "power to act [*puissance d'agir*]," which they call "latitude" (Deleuze and Guattari 1980, 313–14; 1987, 256–57). In other words, the longitude of a body includes the modification of relations of the extensive parts of a body resulting from an encounter, whereas the latitude of a body comprises the affects or the capacities to act and to be acted on of which a body is capable at any one time in an assemblage. What are these acts of which a body is capable? Deleuze and Guattari define affects as "becomings" or capacities to produce emergent effects in entering assemblages (Protevi 2006). These emergent effects will either mesh productively with the affects of the body or clash with them. Meshing emergent effects will augment the power of that body to form other connections within or across assemblages, resulting in joyous affects, while clashing emergent effects will diminish the power of the body to act, producing sad affects.

For Deleuze and Guattari, knowledge of the affects of a body is all-important: "We know nothing about a body until we know what it can do [*ce qu'il peut*], in other words, what its affects are" (Deleuze and Guattari 1980, 314; 1987, 257). Affect is part of their dynamic interactional ontology so that defining bodies in terms of affects or power to act and to undergo is different from reading them in terms of properties of the substantive bodies by which they are arranged in species and genera (314; 257). At this point in their text, Deleuze and Guattari illustrate the way affect is part of the process of assembling by reference to the relation between Little Hans and the horse in Freud's eponymous case study. While we will not do a thematic study of the horse in *A Thousand Plateaus,* we should recall the prevalence of horses (alongside wolves and rats) in the discussions of affect therein: besides the Little Hans case, we also find the becoming-horse of the masochist being submitted to dressage (193; 155) and, of course, the repeated analyses of man–horse assemblages in the Nomadology chapter (the stirrup, the chariot, etc.). We will return to the question of horse–man assemblages in ancient warfare, in particular, to a recent thesis whereby the defeat of the light chariot forces by berserker runners is the crucial factor in the defeat of Bronze Age kingdoms that marks the 1200 BCE collapse, leading to the eventual emergence of the polis–phalanx assemblage (Drews 1993).

Let us stay with the horse to illustrate affect as the capacity to become, to undergo the stresses inherent in forming a particular assemblage. We can begin by noting that in a grouping based on affect, a racehorse, which carries a rider in a race (i.e., the horse enters the racing assemblage), has more in common with a motorcycle than it does with a plow horse. In turn, the plow horse, which pulls a tool gouging the earth (i.e., the horse enters the agricultural assemblage), has more in common with a tractor.[1] This is not to say that what is usually named a "plow horse" or a "tractor" cannot be made to race, just as "racehorses" and "motorcycles" can be made to pull plows. These affects qua changes in the triggers and patterns of their behavior would constitute another becoming or line of flight, one that is different from their usual, statistically normal ("molar") usages. This becoming would constitute their enlistment in assemblages that tap different "machinic phyla" (bio-techno-social fields for the construction of assemblages [Deleuze and Guattari 1987, 406–11]) and "diagrams" (the patterns that direct the construction of assemblages [141]) than the ones into which they are usually recruited. Whether the bodies involved could withstand the stresses they undergo in these idiosyncratic assemblages is a matter of (one would hope careful) experimentation. Such experimentation—establishing the affects of assemblages, the potentials for emergent functionality—is, I believe, one of the senses we can give to Deleuze's at first glance perplexing term *transcendental empiricism* (Deleuze 1994, 56).

To recap, then, Deleuze and Guattari follow Spinoza, defining affect as a body's ability to act and to be acted on, what it can do and what it can undergo. Deleuze and Guattari operationalize the notion of affect as the ability of bodies to form "assemblages" with other bodies, that is, to form emergent functional structures that conserve the heterogeneity of their components. For Deleuze and Guattari, then, affect is physiological, psychological, and machinic: it imbricates the social and the somatic in forming what I've called a *body politic*, which feels its power or potential to act increasing or decreasing as it encounters other bodies politic and forms assemblages with them (or, indeed, fails to do so). In this notion of assemblage as emergent functional structure, that is, a dispersed system that enables focused behavior at the system level as it constrains component action, we find parallels with novel positions in contemporary cognitive science (the "embodied" or "extended" mind schools), which maintain that cognition operates in loops among brain, body, and environment (Clark 2003; Thompson 2007). In noting these parallels, we should point out that Deleuze and Guattari emphasize the affective dimension of assemblages, while the embodied–embedded school focuses on cognition. While we follow Deleuze and Guattari's lead and focus on the affective, we should remember that both affect and cognition are aspects of a single process, *affective cognition*, as the directed action of a living being in its world (Protevi 2009).

In discussing affect, we should note that Deleuze and Guattari regard feeling as the subjective appropriation of affect. An example would be the way pleasure is for them the subjective appropriation of desubjectivizing joyous affect: "pleasure is an affection of a person or a subject; it is the only way for persons to 'find themselves' in the process of desire that exceeds them; pleasures, even the most artificial, are reterritorializations" (Deleuze and Guattari 1987, 156). More generally, "feeling" *(sentiment)* is the subject's self-defense against excessive physiological–emotional changes of the body: "Affect is the active discharge of emotion, the counterattack [*la riposte*], whereas feeling [*le sentiment*] is an always displaced, retarded, resisting emotion" (Deleuze and Guattari 1980, 498; 1987, 400). Deleuze and Guattari's point about affect's extension beyond subjective feeling dovetails with the analysis we will develop of extreme cases of rage and panic as triggering an evacuation of the subject as automatic responses take over; as we will put it, drastic episodes of rage and fear are desubjectivizing. The agent of an action undertaken in a rage or panic state can be said to be the embodied "affect program" (Griffiths 1997) acting independently of the subject. Here we see affect freed from subjective feeling. There can be no complaints about eliminating the first-person perspective in studying these episodes because there is no first-person operative in these cases. Agency and subjectivity are split; affect extends beyond feeling; the body does something, is the agent for an action, in the absence of a subject, as we argued in chapter 2.

Taking up analyses from Protevi (2009), let me give a brief example of research in social psychology that recognizes the ontology of affect in bodies politic. Nisbett and Cohen (1996, 44–45) go below the conscious subject to examine physiological response, demonstrating that white males of the southern United States had markedly greater outputs of cortisol and testosterone in response to insults than a control group of northern white males. They go above the individual subject to examine social policy forms, showing that southern states have looser gun control laws, more lenient laws regarding the use of violence in defense of self and property, and more lenient practices regarding use of violence for social control (domestic violence, corporal punishment in schools, and capital punishment) (57–73). They also offer, though in passing, some speculation as to the role played by slavery in the South in constructing these bodies politic, in which social institutions and somatic affect are intertwined and mutually reinforcing in diachronically developing and intensifying mutual reinforcement. However, as the title of their book indicates, and as is stressed by Jesse Prinz (2012) in his reading of their work, the culture of honor is more related to Scots-Irish herding practices than to farm practices (Prinz 2012, 328). The important point, however, is that it is cultural subjectification practices that instill the triggers and thresholds for the release of hormones and neurotransmitters that ease the way for—though certainly do not determine—the violent actions in question. Just as no one should think the cortex is irrelevant here—there is no bottom-up subsubjective neurophysiological determinism evoked here at the behavioral level—no one should think experience is irrelevant here and turn to a genetic determinism on the ontogenetic level. That is, no one should think that these southern males have a significantly different genetic makeup from other groups of Americans (or better, that any genetic variation is larger within the group than is present between this group and others); the difference in reaction comes from the differences in bodies politic formed by different subjectification practices, that is, the differences in the way social practices have installed cortically mediated triggers and thresholds that actualize the anger potentials we all share due to our common genetic heritage.

Thus, as we have seen, affect is inherently political: bodies are part of an eco-social matrix of other bodies, affecting them and being affected by them. As we will now see, important schools of biological thought accord with this notion of affect as biocultural.

The Biology of Affect

Let us first consider the neuroscience of affect. Again returning to analyses first developed in Protevi (2009), let us focus on rage, as the triggering of this desubjectizing affect was the target of constructions in the geo-bio-techno-affective assemblages of ancient warfare (see also chapter 2). Rage is a basic emotion that is not to be confused with aggression, though it is sometimes at the root of aggressive behavior. A leading neuroscientist investigating rage is Jaak Panksepp (1998), whose *Affective Neuroscience* is a standard textbook in the field. He argues that aggression is wider than anger (187), distinguishing at least two forms of "aggressive circuits" in mammalian brains: predation and rage (188). Predation is based in what Panksepp calls the *seeking* system, which is activated by physiological imbalances, those that can be experienced as hunger, thirst, or sexual need. In predatory hunting, based in seeking, the subject is still operative; there is an experience to hunting, we can experience "what it is like" to hunt. Now we must be careful about too strictly distinguishing predation and rage in the act of killing. Concrete episodes are most often blends of anger and predation; as one expert puts it, "real-life encounters tend to yield eclectic admixtures, composites of goal and rage, purpose and hate, reason and feeling, rationality and irrationality. Instrumental and hostile violence are not only *kinds* of violence, but also violence qualities or *components*" (Toch 1992, 1–2; emphasis original).

Although in many cases, we find composites of brute rage and purposeful predation, we can isolate, at least theoretically, the pure state or blind rage in which the subject drops out. We take the Viking berserker rage as a prototype, a particularly intense expression of the underlying neurological rage circuits that evacuates subjectivity and results in a sort of killing frenzy without conscious control. The notion of a blind, desubjectified, rage is confirmed by Panksepp's (1998, 196) analysis of the "hierarchical" architecture of the neural circuits involved: "the core of the RAGE system runs from the medial amygdaloid areas downward, largely via the stria terminalis to the medial hypothalamus, and from there to specific locations with the PAG [periaqueductal gray] of the midbrain. This system is organized hierarchically, meaning that aggression evoked from the amygdala is critically dependent on the lower regions, while aggression from lower sites does not depend critically on the integrity of the higher areas."

We must admit that there are huge issues here with the relation of Panksepp's anatomical focus on specific circuits and neurodynamic approaches that stress that the activity of multiple brain regions is involved in the activation of any one brain function; this anatomy versus dynamics relation must itself be seen in the historical context of the perennial localist versus globalist debate. We are in no position to intervene in these most complex issues, but we should note that Panksepp's notion of hierarchical circuits does allow for the possibility that "higher areas provide subtle refinements to the orchestration that is elaborated in the PAG of the mesencephalon [midbrain]. For instance, various irritating perceptions probably get transmitted into the system via thalamic and cortical inputs to the medial amygdala" (196–97). While these "irritating perceptions" may simply stoke the system to ever-greater heights of rage, we do need to allow that, in some cases, conscious control can reassert itself.[2] Nonetheless, Panksepp's basic approach, as well as the volumes of warrior testimony about the berserker rage (Shay [1994] being only the tip of the iceberg), licenses our description of the pure berserker rage as blind and desubjectified.

Now it is not that the Viking culture somehow presented simply a stage for the playing out of these neurological circuits. To provoke the berserker rage, the Vikings, through a variety of training practices embedded in their customs, distributed traits for triggering the berserker process throughout their population. Presumably, they underwent an evolutionary process in which success in raiding undertaken in the berserker frenzy provided a selection pressure for isolating and improving these practices. (We will return to the question of cultural evolution later; for the moment, please note that I am not implying that genes were the target of that selection pressure.) In other words, the Vikings explored the biosocial machinic phylum for rage triggers in their military assemblages. While one researcher cites possible mushroom ingestion as a contributing factor (Fabing 1956), we will nonetheless rather focus on the role of dance and song in triggering the berserker state. In his important work *Keeping Together in Time: Dance and Drill in Human History,* the noted historian William McNeill (1995, 102) notes that "war dances" produced a "heightened excitement" that contributed to the "reckless attacks" of the Viking berserkers (see also Speidel 2002, 276).

There is no denying that the social meaning of blind rages differs across cultures—how they are interpreted by others and by self after waking up—as do their manifold triggers and thresholds (Mallon and Stich 2000). But I think it is important to rescue a minimal notion of human nature from extreme social constructivism and to hold that rage episodes are individuations of a multiplicity encompassing variation in genetic inheritances, environmental input in the form of subjectification practices, and developmental plasticity. Even with all that variation, there is still remarkable similarity in what a full rage episode looks like (the differenciations or types of rage states resulting from those individuations are tightly clumped into a broadly recognizable "rage pattern," in other words), though how much it takes to get there, and what the intermediate anger episodes look like ("emotion scripts" [Parkinson, Fischer, and Manstead 2005]), can differ widely. Even James Averill, a leading social constructivist when it comes to emotion, relates "running amok" in Southeast Asian societies to Viking berserker rages. Averill (1982, 59; emphasis original) writes, "Aggressive frenzies are, of course, found in many different cultures (e.g., the *'berserk'* reaction attributed to old Norse warriors), but amok is probably the most studied of these syndromes." It is the very *commonality* of "aggressive frenzies" that we are after in our notion of a broadly applicable "rage pattern." To recall briefly the analyses of chapter 2, I propose that in extreme cases of rage, a modular rage agent replaces the self-conscious subject with what is called an *affect program,* that is, an emotional response that is "complex, coordinated, and automated . . . unfold[ing] in this coordinated fashion without the need for conscious direction" (Griffiths 1997, 77). Affect programs such as rage (panic is another example) are more than reflexes, but they are triggered well before any cortical processing can take place (though later cortical appraisals can dampen or accelerate the affect program).

Placing the short-term behavioral notion of rage agent in its full ontogenetic and phylogenetic context, we cannot think of bodies politic as mere input–output machines passively patterned by their environment (a discredited social constructivism) or passively programmed by their genes (an equally discredited genetic determinism). We thus turn to an important school of thought in contemporary critical biology, DST, which is taken from the writings of Richard Lewontin (2002), Susan Oyama (2000), Paul Griffiths and Russell Gray (1997; 2001; 2004; 2005), and others (Oyama, Griffiths, and Gray 2001). With the help of this new critical biology, we can see the body politic as developmentally plastic, that is, as neither a simple blank slate nor a determined mechanism. Instead, the body politic is biologically open to the subjectification practices it undergoes in its cultural embedding, practices that work with the broad contours provided by the genetic contribution to development to install culturally variant triggers and thresholds to the basic patterns that are our common heritage. Griffiths uses the example of fear to make this point, but the same holds for the basic emotion of rage we discussed earlier. "The empirical evidence suggests that in humans the actual fear response—the output side of fear—is an outcome of very coarse-grained selection, since it responds to danger of all kinds. The emotional appraisal mechanism for fear— the input side—seems to have been shaped by a combination of very fine-grained selection, since it is primed to respond to crude snake-like gestalts, and selection for developmental plasticity, since very few stimuli elicit fear without relevant experience" (Griffiths 2007, 204).

We cannot enter the details of the controversy surrounding the notion of multiple levels of selection here, but we can at least sketch the main issues surrounding the notion of group selection, which plays a key role in any notion of biocultural evolution (Sober and Wilson 1998; Sterelny and Griffiths 1999; Jablonka and Lamb 2005; Joyce 2006). In considering the notion of group selection, we find two main issues: emergence and altruism. If groups can be emergent individuals, then groups can also be vehicles for selection. For example, groups that cooperate better may have outreproduced those that did not. The crucial question is the replicator, the ultimate target of the selection pressures: again, for our purposes, the unit of selection is not the gene or the meme (a discrete unit of cultural "information") but the set of practices for forming bodies politic. With the cooperation necessary for group selection, we must discuss the notion of *altruism* or, more precisely, the seeming paradox of *fitness-sacrificing behavior*. It would seem that natural selection would weed out dispositions leading to behaviors that sacrifice individual fitness (defined, as always, as the frequency of reproduction). The famous answer that seemed to put paid to the notion of group selection came in the concept of *kin selection*. The idea here is that if you sacrifice yourself for kin, at least part of your genotype, the "altruistic" part that determines or at least influences self-sacrifice and that is (probably) shared with that kin, is passed on. But again, all the preceding discussion operates at the genetic level. We will claim that the ultimate target of selection pressure in group selection for the production of bodies politic is the set of social practices reliably producing a certain trait by working with our genetic heritage. This need not have any implications for genetic fitness sacrificing in group selection, if we restrict ourselves to bodies politic and the social practices for promoting behavior leading to increased group fitness. In other words, we are concerned with the variable cultural setting of triggers and thresholds for minimally genetically guided basic patterns.

The important thing for our purposes here is the emphasis DST places on the life cycle, developmental plasticity, and environmental co-constitution. In following these thinkers, we can replace the controversial term *innate* with (the admittedly equally controversial) "reliably produced given certain environmental factors." In so doing, we have room to analyze differential patterns in societies that bring forth important differences from common endowments. In other words, we don't genetically inherit a subject, but we do inherit the potential to develop a subject when it is called forth by cultural practices. It is precisely the various types of subjects called forth (the distribution of cognitive and affective patterns, thresholds, and triggers in a given population) that are to be analyzed in the study of the history of affect.

The History of Affect

DST enables us to explore the biocultural dimension of bodies politic by thematizing extrasomatic inheritance as whatever is reliably reproduced in the next life cycle. Thus, with humans, we're into the realm of biocultural evolution, with all its complexity and debates.[3] Again, the unit of selection here is not purely and simply genetic (indeed, for the most part, genes are unaffected by cultural evolution, the classical instances of lactose tolerance and sickle cell anemia notwithstanding). Rather, the unit of selection should be seen as sets of cultural practices, thought of in terms of their ability to produce affective cognitive structures (tendencies to react to categories of events) by tinkering with broadly genetically guided neuroendocrine developmental processes such as in the case of the Viking's having honed in on rage-triggering practices.[4]

Regarding historically formed and culturally variable affective cognition, I work with Damasio's framework for the most part (Damasio 1994; 1999; 2003). Brain and body communicate neurologically and chemically in forming "somatic markers," which correlate or tag changes in the characteristic profile of body processes with environmental encounters, that is, changes in the characteristic profile of body–world interactions, which provoke them. Somatic markers are formed via a complex process involving brain–body–environment interaction, in which various sectors of the brain receive signals from multiple sources: from the body, from brain maps of body sectors, and from its own internal self-monitoring sectors. Thus the brain synthesizes how the world is changing (sensory input, which is only a modulation of ongoing processes), how the body is being affected by the world's changing (proprioception or *somatic mapping*, again, a modulation of ongoing processes), and how the brain's endogenous dynamics are changing (modulation of ongoing internal neurological traffic or *meta-representations*). This synthesis sets up the capacity to experience a feeling of how the body would be affected were it to perform a certain action and hence be affected in turn by the world (off-line imaging, that is, modulation of the ongoing stream of "somatic markers"). I cannot detail the argument here, but a neurodynamic reading of Damasio's framework is broadly consonant with the Deleuzian emphasis on differential relations, that is, the linkage of rates of change of neural firing patterns, and on their integration at certain critical thresholds, resulting in "resonant cell assemblies" (Varela 1995) or their equivalent (Kelso 1995; Edelman and Tononi 2000). The key is that the history of bodily experience is what sets up a somatic marker profile; in other words, the affective cognition profile of bodies politic is embodied and historical.[5]

With this background, we see that much of the controversy surrounding cultural evolution is due to the assumption that information transfer is the target for investigation (Runciman 2005a, 13). But the notions of *meme* and *information transfer* founder on DST's critique—it is not a formed unit of information that we're after but a process of guiding the production of dispositions to form somatic markers in particularly culturally informed ways. We have to think of ourselves as biocultural, with minimal genetically guided psychological modularity (reliably reproduced across cultures) and with a great deal of plasticity allowing for biocultural variance in forming our intuitions (Wexler 2006). In other words, we have to study political physiology, defined as the study of the production of the variance in affective cognitive triggers and thresholds in bodies politic, based on some minimally shared basic patterns.

All this means that we cannot assume an abstract affective cognitive subject but rather have to investigate the history of affect. But, the objection might go, don't we thereby risk leaving philosophy and entering historical anthropology? The answer is that we only leave philosophy to enter history if we've surreptitiously defined philosophy ahead of time as ahistorical. Well, then, don't we leave philosophy and enter psychology? Only if we've defined philosophy as concerned solely with universal structures of affective cognition. But that's the nub of the argument: the abstraction needed to reach the universally "human" (as opposed to the historically variant) is at heart antibiological. Our biology makes humans essentially open to our cultural imprinting; our nature is to be so open to our nurture that it becomes second nature. But just saying that is typological thinking, concerned with "the" (universal) human realm; we need to bring concrete biological thought into philosophy. It is the variations in and across populations that are real; the type is an abstraction.

Having said all that, we must be clear that we are targeting variation in the subjectification practices producing variable triggers and thresholds of shared basic patterns. Now almost all of us reliably develop a set of basic emotions (rage, sadness, joy, fear, distaste), ones that we share with a good number of reasonably complex mammals (Panksepp 1998). Many of us also have robust and reliable prosocial emotions (fairness, gratitude, punishment—shame and guilt are controversial cases) we share with primates, given certain basic and very widespread socializing inputs (de Waal 2006; Joyce 2006). Although some cultural practices can try to expand the reach of prosocial emotions to all humans or even to all sentient creatures (with all sorts of stops in between), in many sets of cultural practices, these prosocial emotions are partial and local (Hume's starting point in talking about the "moral sentiments"). Why is the partiality of prosocial emotions a default setting for sets of biocultural practices? One hypothesis is that war has been a selection pressure in biocultural evolution, operating at the level of group selection and producing very strong in-group versus out-group distinctions and very strong rewards and punishments for in-group conformity (e.g., Bowles and Gintis 2003).

There are difficult issues here concerning group selection and the unit of selection (Dawson 1999), but even if we can avoid the genetic level and focus on group selection for sets of social practices producing prosocial behaviors, we must still take into account a bitter controversy in anthropology about the alleged universality of warfare in human evolution and history.[6] There are three elements to consider here: the biological, the archaeological, and the ethnographic. (1) Regarding the biological, an important first step is to distinguish human war from chimpanzee male coalition and aggressive hierarchy, to which it is assimilated in the "humans as killer apes" hypothesis (Peterson and Wrangham 1997). Criticizing the killer apes hypothesis, several researchers point out that we are just as genetically related to bonobos, who are behaviorally very different from chimpanzees (de Waal 2006, 73; Fry 2007; see also Ferguson 2008). (2) Proponents of universal war often point to findings of crushed skulls and the like in the archaeological record (Keeley 1997). Critics reply that some of the claims of war-damaged skulls are more plausibly accounted for by animal attacks (Fry 2007, 43). (3) Finally, we must couple the archaeological record with the current ethnographic record. But to do that, we must distinguish smaller and less internally differentiated forager bands from more internally complex hunter-gatherer tribes with chiefs. Though the critics of the universal war thesis admit that the former social groups had individual-level murder and revenge killing and even group "executions" of murderous individuals, they deny that they had the "logic of social substitutability" that enables warfare as anonymous group-level conflict in which any member of the opposing group is fair game (Kelly 2000; Fry 2007). The critics of universal war also remind us of the need to look at current tribal warfare in the context of Western contact and subsequent territorial constriction and / or rivalry over trading rights (Sponsel 2000; Ferguson 1995).

The question would be how much war was needed to form an effective selection pressure for strong in-group identification and hence partiality of prosocial emotions? Richerson and Boyd (2005, 209–10) argue that between-group imitation can also be a factor in spreading cultural variants. Richerson and Boyd cite the example of early Christianity, where the selection pressure for subjectification practices of "brotherhood" and hence care for the poor and sick was the high rate of epidemics in the Roman Empire. So war need not be the only selection pressure, nor does group destruction and assimilation of losers have to be the only means of transmitting cultural variants. We will assume in the following section that we haven't had time for selection pressures on genes with regard to warfare (Dawson 1999), but we have had time for selection pressures on biocultural subjectification practices relative to warfare, for example, how to entrain a marching phalanx versus how to trigger berserker rage.

If war was a selection pressure on group subjectification practices for forming different bodies politic, we have to consider the history of warfare. With complex tribal warfare, you get loose groups of warriors with charismatic leaders (Clastres 1989). Virtually all the males of the tribe take part in this type of warfare; in other words, there is no professional warrior class–caste, except in certain rare cases. Fry's (2007) argument is that the Chagnon–Clastres school, which focuses on the Yanomami as prototypical "primitive" warriors, picked complex horticultural hunter-gatherers and missed the even more basic simple foragers, who actually represented the vastly most occurrent social organization in human history. But our critique lies elsewhere because we are not talking about genes but about biocultural evolution, about group selection of affective practices. So to investigate the role of warfare in the history of affect, we do not have to claim that warfare is in our genes; we need only investigate the geo-bio-affective group subjectification practices, once warfare is widespread with either internally differentiated tribes or with State societies (Lawler 2012b; Bowles 2012).[7] Actually, we can argue that the spread of warfare coincides with agricultural States, even in complex tribal, "primitive" societies, if we accept Deleuze and Guattari's claim that such tribes have always had States on their horizon (both immanently, as that which is warded off, and externally, as that which is fought against; again, see Ferguson [1995] for a political history of Yanomami warfare).

It is certainly an anthropological staple that forager band egalitarianism changes with agriculture and class society. We need not enter Deleuze and Guattari's "anti-evolution" argument and the notion of the always-existent *Urstaat* (Deleuze and Guattari 1987, 357–61), though we can note some fascinating new research that broadly supports their claim, derived from Jane Jacobs (1970), of the urban origins of agriculture (Balter 1998; but see the nuanced multiple-origins account of Pringle 1998). Consider thus the situation in Homer: we see vast differences between the affective structures of the warriors (bravery), the peasants (docility) who support them, the artisans who supply the arms (diligence), and the bards who sing their praises and who thus reinforce the affective structures of the warriors: the feeling that your name will live on if you perform bravely is very important. Thus, here, the selection pressure is for sets of biocultural practices producing specialized affective structures relative to position in society, that is, relative to their contribution to the effectiveness of wars fought by that society. Once again, our concern is with the biocultural production of bodies politic, which tries to reliably produce bioaffective states. As we argued in chapter 1, and will explore further later, the triumph of hoplite phalanx warfare marks a shift in biocultural production. Compare Aristotle's notion of courage with what the Homeric warriors understood as courage. For Aristotle, courage means staying in the phalanx with your mates so that charging ahead rashly is as much a fault as cowardly retreat (*Nicomachean Ethics* 2.8.1108b20–27; 3.7.1115a30; b25–30; cf. Hanson 1989, 168); for the Homeric heroes, charging ahead rashly is all there is.

Music and Ancient Warfare

We have now set up our research question. For a final preparation for our study of the role of music in the affective assemblages that trigger berserker rage or that entrain the phalanx, let us consider recent research that has proposed studying music in the context of biocultural evolution. The leading researcher of the Cambridge school of thought in this area, Ian Cross, argues that music is ancient and universal for humans—so ancient that here we can consider a significant genetic component. Against two recent claims, Cross holds that Steven Pinker is wrong about music being only a spandrel and that Geoffrey Miller is wrong about it being due to sexual selection because they both think of music as contemporary Western music experience, as "patterned sound employed primarily for hedonic ends, whose production constitutes a specialised, commodified and technologised activity" (Cross 2003, 13). Another member of the Cambridge school of thought on music and biocultural evolution, John Bispham, puts the contrary position as clearly as possible: "music is a culturally constructed phenomenon built upon universal biologically determined foundations" (Bispham 2004).

Now we must be clear that studying music in an evolutionary framework does not yield a simple adaptive story. Rather, it seems that various "proto-musical" capacities evolved separately and later were stitched together to yield human musical capacities. Bispham (2006, 125) proposes that musical rhythmic behavior "be viewed as a constellation of concurrently operating, hierarchically organized, subskills including general timing abilities, smooth and ballistic movement (periodic and nonperiodic), the perception of pulse, a coupling of action and perception, and error correction mechanisms"; all of these "subskills share overlapping internal oscillatory mechanisms." These various capacities should be seen as "grounded in, and as having exaptively evolved from, fundamental kinesthetic abilities and modes of perceiving temporally organized events" (125). In sum, Bispham is against a straight-line evolutionary story: "complex behaviors such as music evolved in a mosaic fashion, with individual components emerging or evolving independently or for independent reasons at times, and / or reforming with other components at other times" (126). This doesn't mean that any one mechanism wasn't selected for, just not the full combination as such, until much later, after independent evolution of the components. The evolutionary pressures that have shaped the fundamentally rhythmic and social aspects of our being lead Cross to claim that "infants appear to be primed for music"; in support of this, he cites important studies on rhythmic mother–infant interactions, which are crucial for "primary intersubjectivity," "emotional regulation," and "emotional bonding" (Cross 2003, 16; citing Trevarthen 1999; Dissanayake 2000). In the same vein, Bispham (2006, 125) classifies Dissanayake (2000) as looking for "the adaptive strength of rhythm and entrainment in the course of human evolution with reference to mother–infant interaction."

These early building blocks of musicality must come together to form our uniquely human rhythmic capacities. Ian Cross (2003, 18), for his part, insists that the cultural evolution of music cannot be about "memes," which are discrete and consist in "information transfer." The idea has to be that music is involved in the development of bodies politic. For our purposes, music is a powerful way of searching the machinic phylum for biosocial assemblage formation to draw out practices forming bodies politic that can contribute to group functionality. The key here is to see "interpersonal musical entrainment" as the uniquely human musical capacity. What distinguishes human music from birdsong is that, so often, our music is a group activity involving changes in response to changes by others (Bispham 2004). Thus a key capacity for investigation is entrainment, or group movement with the same pulse, which plays a major role in Bispham's analysis; entrainment is based on "internal oscillatory mechanisms [that] are attuned to external cues allowing us to build expectations for the timing of future events . . . and to interact efficiently with the environment" (Bispham 2006, 128). Because there are internal oscillatory mechanisms in a variety of domains of human behavior and cognition, this suggests that "entrainment in music constitutes an evolutionary exaptation of more generally functional mechanisms for future-directed attending to temporally structured events" (128). Bispham pushes the analysis as far as to entertain the notion that "interpersonal entrainment is the key rhythmic feature in [all] human interactions," both musical and nonmusical. Such interpersonal entrainment ranges from "loose, subconscious use of pulse as a framework for interpersonal/ turn-taking interactions in, for example, mother–infant or linguistic interactions" to "a strict adherence to pulse (groove) in group behavior and synchronicity of output where participants are aware of the pulse framework and desire to maintain a degree of temporal stability and group-coordination (e.g., music and dance)" (129). However, Bispham claims that it is probably the case that the former precedes the latter ontogenetically (and perhaps phylogenetically) and is less complex psychologically and physiologically. So in his search for what is unique about human musical rhythm, Bispham focuses on musical pulse and period correction as the keys here; we will deal only with the first of these.

Regarding musical pulse, we have to remember that for almost all of human history, music has had to have been danceable, which sets up its capacity for group bonding (Dissanayake 2000; McNeill 1995). How does danceability come about? Bispham (2006, 129) points first to "internal periodic oscillatory mechanisms overlapping with motor-coordination." The key for us is his conclusion that this provides "a mechanism to affect and regulate levels of physiological arousal." In other words, music allows groups to get on the same emotional wavelength: "musical pulse is functional in regulating emotions and motivational states by means of affecting states of action-readiness" (131). It is important to stress that in an evolutionary perspective, music is regulatory rather than merely expressive: music is "functional in regulating emotions and in communicating strategies for the regulation of emotion rather than as raw emotional expression per se" (131). Most importantly for our purposes, Bispham notes that such functional affective regulation by means of group music includes "military arousal" (130).

With this background, let us now narrow the focus to music and ancient warfare. First, we can distance ourselves from the approach taken by Hagen and Bryant (2003), whose investigation of music as a "coalition signaling system" focuses on a purely genetic level of selection. They believe themselves licensed to do so because they assume the ancient universality of warfare, but we saw earlier how this is based on highly questionable arguments. We will instead focus on much more recent events, in particular, on the differences between the biosocial assemblages of the berserker rage triggered by the war dance and the entrainment of the phalanx by cadence marching. To set the stage, let us reconstruct the commonly accepted chronology of military assemblages in the ancient Near East. The separation of military and priestly power, according to William McNeill (1995, 105), had "a distinct historical origin in ancient Sumer, when, in the language of their epic of creation, 'kingship came down from the gods' to challenge priestly management of society around 3000 BCE. The rise of kingship in Mesopotamia, as it happens, was also connected with the earliest known manifestation of the tamer version of war dances—close-order drill." Reading the epic of Gilgamesh as recording the establishment of kingship from pressures of barbarian raiding and fighting between adjacent cities over water rights, McNeill sees evidence of close-order infantry spearmen about 2450 BCE. Now McNeill holds that Sumerian military–political independence from priestly–religious authority was "exceptional" and due to biosocial entrainment: "I suggest that the psychological impact of drill may well have been critical in keeping the military–political structure of ancient Sumer independent of, and sometimes at odds with, priestly–religious authority" (107). To account for McNeill's observation, we can speculate that the loyalty to military groups evoked by drill-induced entrainment was tightly focused and controlled by the charismatic figure of the leader to whom some of the affect was attached, as opposed to the looser focus of the affect induced by religious ceremonies.[8]

To return from our speculation to the realm of historical evidence, the key change in military tactics in the succeeding time period is the replacement of early infantry by light chariots as mobile archery platforms, the dominant form of military assemblage in the Bronze Age kingdoms (McNeill 1995, 108–9; Drews 1993; Ferrill 1997). After the 1200 BCE collapse (on which we will focus shortly), around 875 BCE, we see archers directly on horseback, as riders, whose ability "to concentrate superior force at any given spot, almost at will" (McNeill 1995, 109), is sure to remind all readers of Deleuze and Guattari of nomad tactics, the ability to hold an open field, to move with "intensive speed," and so on (Deleuze and Guattari 1987, 381). These ninth-century riders are the "zoosexual" horsemen noted by Deleuze and Guattari, whose relation to the succeeding form of the phalanx with the rise of the polis in the eighth and seventh centuries BCE is exceedingly complex and beyond the scope of this chapter to explore fully (Detienne 1968).

Now whatever the precise details of the emergence of the phalanx, it is clear that marching played a major role in its purest expression: with the Spartans (Lazenby 1985). A number of preliminary issues must be discussed here, none of which we can treat in depth, but all of which we can at least note. First of all, the Spartans were the only polis to devote much time to training for warfare, a point of pride for the Athenians, who thought themselves better balanced in their lives (Hanson 1989, 31; see also Pritchett 1974, 214). Second, I find myself compelled once again, at the risk of fatiguing my readers, to insist that the unit of analysis for us in discussing phalanx warfare must be the set of biosocial practices for directing the development of the affective structures of bodies politic, not the all-too-cognitivist and/ or hylomorphic notion of "bundles of information or instructions" (cognitivist because of "information" and hylomorphic because of "instructions") allowing for the "exosomatic" transmittal of "sets of norms, values and beliefs" (Runciman 1998, 734). With all that in mind, a commonly cited passage from Plutarch confirms the internal emotional bonding and the intimidating effect on opponents that Spartans' musical march into battle had: "And when at last they were drawn up in battle array and the enemy was at hand, the king... ordered the pipers to pipe the strains of the hymn to Castor; then he himself led off in a marching paean, and it was a sight equally grand and terrifying when they marched in step with the rhythm of the flute, without any gap in their line of battle, and with no confusion in their souls, but calmly and cheerfully moving with the strains of their hymn into the deadly fight. Neither fear nor excessive fury is likely to possess men so disposed, but rather a firm purpose full of hope and courage, believing as they do that Heaven is their ally" (*Lycurgus,* 21–22).

Let us now turn from the historically well-documented affective aspect of the phalanx to speculate as to the affective assemblages at work in the 1200 BCE collapse. Though Deleuze and Guattari follow Detienne in discussing the distinction of the Dark Age riders and the classical age phalanx, there is an earlier confrontation of military assemblages that sets the stage for that transformation. Our major source here will be Drews (1993). Drews proposes a military explanation for the 1200 BCE collapse. He first eliminates the standard explanations: earthquakes, migrations, ironworking, systems collapse, and raiders. Systems collapse might be a condition of weakness of kingdoms, but it cannot explain the physical destruction of so many palaces right around 1200 BCE because it is a "process and structure" explanation that cannot deal with the "event" of the destruction (88). While systems collapse is often then coupled with the earthquake or migration hypothesis to explain the "incidental" event of palace destruction, Drews insists that the collapse cannot be seen as simply "internal development—the consequence of deterioration in internal systems"; you need to look to the agency of the attackers and the military weakness to which they were responding (89). Conversely, for Drews the raiders hypothesis is correct but incomplete (91). The key question is, why did the raiders become so suddenly successful at 1200 BCE when before, they were only a nuisance (93)? Drews's innovation comes from examining the makeup of ancient Near Eastern imperial armies. Here we find that light chariots were the main weapon system, as mobile archery platforms. As successful as the light chariots were, they became vulnerable to "a new kind of infantry" (97). These new foot soldiers "used weapons and guerrilla tactics that were characteristic of barbarian hill people but had never been tried en masse in the plains and against the centers of the Late Bronze Age kingdoms" (97).

Acknowledging the speculative character of Drews's work on the level of history (97–98) and our own biocultural and neurological speculation, we focus on his claim that barbarian troops "awoke to a truth that had been with them for some time: the chariot-based forces on which the Great Kingdoms relied could be overwhelmed by swarming infantries" (104). These phrases will draw the notice of every reader of *A Thousand Plateaus*: *swarming* is the key term in the nomad creation of smooth space, or better, the smoothing force of the nomad war machine, which can pop up anywhere in the field in the manner of a "vortex" (Deleuze and Guattari 1987, 381). Let us continue: not only is the light chariot an assemblage in the Deleuzoguattarian sense, bringing horses, chariot, driver, archer, and bow into an emergent functional structure (105–6), but we also see Drews supporting Deleuze and Guattari's thesis of the precedence of the "social machine" over technology. Only with changes in the social structure could the assemblage be fully integrated: "the chariot became militarily significant when it was combined with another intricate artifact, the composite bow, which also had been known for a long time but had until then been a luxury reserved for kings or the very rich" (Drews 1993, 105–6).

The question of Homer's mistakes in portraying the use of chariots as "battle taxis" is fairly well known and is deftly handled by Drews: "I would suggest, then, that Homer was basically ignorant of chariot warfare because the heroic tradition originated in a society of infantry-men, in which the chariot was indeed nothing more than a prestige vehicle" (117). So much is familiar, but here's the twist that establishes Drews's importance for our interests: "Homer's Achaeans were . . . responsible for putting an end to chariot warfare and to the domination of the horse-tamers [cf. Hector's sobriquet]. They were, that is to say, infantrymen of the new type—fleet of foot, skilled with the javelin or throwing spear, and also carrying long swords—who spelled the doom of the great chariot forces of the Late Bronze Age" (118). Drews insists that before the 1200 BCE catastrophe, the foot soldiers supported the chariots, performing the hand-to-hand combat in plains battles (141–42) and doing the fighting in the hills, where chariots couldn't go (147). We see here the outlines of the geographical dimension of the ancient warfare multiplicity, which Drews confirms in his analysis of the fighting styles of the skirmisher-runners, who were barbarian recruits from those very hill regions: "Mobility rather than solidarity was essential" (152). In another aspect that will delight readers of *A Thousand Plateaus,* but that we cannot pursue, Drews indicates the becoming-animal of the prototypical runner, who was "ferocious in his horned or feathered helmet" (152; see also Speidel 2002). A final geo-bio-social ontogenetic element deserves recognition, that is, the differentials in physical conditioning between the barbarian hill runner and the plains-dwelling peasant: "service as skirmishers was undoubtedly hazardous and demanding and must have required a great deal more stamina, skill, recklessness, and perhaps ferocity than could be found in the typical resident of Ugarit, Messenia, or Memphis" (157).[9]

The key question for us, turning now to individuations on the behavioral level, must be, what is the affect of Achilles the fleet footed? We all know the answer: speed and rage at close range. As much as anyone in our culture, Achilles is the prototype of the berserker rage (Harris 2001; Shay 1994). Drews is clear that the runner-raiders fought in the style that Deleuze and Guattari associate with the steppe nomads: "With a long sword as his primary weapon for hand-to-hand warfare, the raider required an 'open' space, in which his agility and fleetness could be exploited" (Drews 1993, 210). Now, of course, we know that Homer paints Hector as having the same affect, but according to Drews's novel interpretation, Troy itself is a horse (chariot) city that is sacked by the raiders (211). Drews's point is that in Homer's portrayal, Achilles and the Greeks fight as did the barbarian hill people who had been incorporated into the complex armies of the Bronze Age kingdoms but then discovered that they could defeat chariots. Whatever the worth of these speculations as literary criticism, the important point for us is the affect of the runners from the hills who made up the skirmishers of the imperial armies. Here we turn to a fascinating article by Michael Speidel (2002, 258), who makes the argument that the hill people were Indo-European, fighting in berserker style. Despite his wonderful descriptions of the berserkers, Speidel unfortunately talks about the "berserk mind" (259), where we must insist that *body politic* is a much better term. Speidel does, however, use a great, if slightly pleonastic, term for their tactics, one that is again sure to draw the attention of readers of *A Thousand Plateaus*: the berserkers fought as in a "swirling whirlwind" (259).

To focus our concern on the behavioral individuation of acts of bodies politic, we ask, what were the techniques to bring out the berserker rage? Music is primary among these triggers. Speidel tells us that "to do deeds of berserk daring, one had to be raging mad. . . . Shouting and singing were ways to raise such rage. Early Greek and Roman warriors screeched like flocks of raucous birds—a mark of manhood" (Speidel 2002, 273, citing *Iliad* 3.2–6 and *Aeneid* 7.705). Speidel notes the cross-cultural effectiveness of the war song in provoking the berserker rage: "with a song of thunder and wind, the young Marut warriors of the Rig Veda awakened Indra's prowess. Husky Thracian, Celtic and Germanic war songs, like crashing waves, heartened warriors" (273). We should also connect Speidel's next point with the research on musical pulse and the insistence on dance we found among the Cambridge school researchers, Cross and Bispham. Speidel writes, "Dance emboldened even more. Not only Tukulti-Ninurta's berserks danced on the battlefield: Vedic Indians did the same. . . . Dances, though done by all the early warriors, mattered particularly to berserks as they fanned their fury" (273–74). Speidel cites only "adrenaline levels" (276) as the physiological component of the berserker rage, but we can do better with our understanding of Panksepp and biocultural triggering.

I am going to speculate that these dances and songs were "rhythms," in Deleuze and Guattari's technical sense, rather than "cadences" like phalanx marches (Deleuze and Guattari 1987, 313; see Bogue 2003; Turetsky 2004; Grosz 2008). Now, given the complexity of the conceptual network Deleuze and Guattari use to discuss living systems in *A Thousand Plateaus,* we can only sketch some of the relations among the key terms: *milieus* and *codes,* and *strata* and *territories.* We begin with *milieu,* which is a vibratory, rhythmic, and coded material field (313) for bodies (strata) and territories (assemblages). Heterogeneous milieus are "drawn" by rhythms from chaos, while territories form between ever-shifting milieus. Now milieus are coded—the "code" is the repetition of elements such that milieus are a "block of space-time constituted by the periodic repetition of the component"—but the rhythm is always shifting in "transcoding" (313). Thus *rhythm* is the difference between one code and another: "there is rhythm whenever there is a transcoded passage from one milieu to another, a communication of milieus, coordination between heterogeneous space-times" (313). Deleuze and Guattari's rhythm is differential: "rhythm is critical; it ties together critical moments" (313). *Critical* here means a threshold in a differential relation, a singularity in the linked rates of change of a living system in its ecological niche.

Looking now to link all three levels of individuation of bodies politic—the phylogenetic (though remembering that the units of selection here are sets of subjectification practices, not the purely genetic), the ontogenetic, and the behavioral—we can speculate that through a coevolutionary process with success in warfare as a selection pressure, the barbarian hill peoples (and others who searched the same "machinic phylum") experimented with war dances and songs to hit on critical points in provoking neural firing patterns that triggered evolutionarily embedded rage circuits or "affect programs," as Griffiths (1997) calls them. Indeed, Panksepp (1998, 197) gives us a clue as to why dance and song were the elements of experimentation: "[Certain brain] areas presumably code the affective content of certain irritations, including vocalizations, and may give specific sounds direct access to RAGE circuitry." Along with the angry tone of the war cry (and here elements of autoaffection must be taken into account—you can participate in an escalating affective episode by your own efforts, as we all know, just as you can calm yourself down with some deep breaths), the exertions of the dance help sensitize the system, that is, lower the threshold for the triggering of the rage episode: "increased activity in baroreceptors of the carotid arteries monitors levels of blood pressure and can facilitate the sensitivity of RAGE circuitry" (Panksepp 1998, 198). We can only speculate as to the neurodynamics and intermodal processing of the auditory and proprioceptive sensations of the dance–song, but we can see here an intermeshing of differential multiplicities, that is, in Deleuze and Guattari's terms, rhythm as "critical."

Thus we can see that the behavioral expression of our ontogenetically crafted triggers for evolutionarily preserved potentials for rages depends on the correct combination of many different layers of events, which the cultural evolutionary process of adjusting the war dance and war song to the triggering of such rages set about exploring. Speidel himself seems to sense that there must have been an imbrication of the social and somatic in the berserker rage, writing that similarities in berserker style, if not due to contact, "must be due to human traits common to the structure and functioning of all warrior societies" (286).

With all that said, we can nonetheless not draw too sharp a distinction between berserker rage and phalanx fighting.[10] To be sure, Speidel (2002, 279) contrasts the "mindlessness" of berserkers and the "disciplined fighting" of Greeks and Romans. But the phalanx doesn't just march. After the clash, we find the chaotic melee, in which some form of rage was certainly called on. One account understatedly questions whether the soldiers in the melee were "rational" and speculates that here the soldiers were on "automatic pilot" (Hanson 1989, 159; scare quotes in original). We will read this "automatic pilot" as a desubjectivizing rage state. Conversely, however chaotic the melee, it was still a clash of phalanxes and thus required both discipline and rage. We are in no position to do more than speculate as to the means by which such balance was achieved, but we are almost irresistibly tempted to use the well-worn Apollo and Dionysus trope about the Greek phalanx warrior, who "was thus asked to accomplish two difficult and almost mutually exclusive tasks: to unleash a wild fury in the initial crash, and then to maintain complete mastery of this savagery, to guide each step into the enemy columns with complete discipline" (Hanson 1989, 169).

PART II
Cognitive Science: Brain and Body

⇒ 4 ⇐

Dynamic Interactionism

Here we begin part II of the book, bringing Deleuze to bear on the cognitive and affective sciences. These sciences have benefited in the last twenty years from a rethinking of the long-dominant computer model of the mind espoused by the standard approaches of computationalism and connectionism. The development of this alternative, often named the *embodied mind* approach or the *4EA* approach (embodied, embedded, enactive, extended, affective), has relied on a trio of classical twentieth-century phenomenologists for its philosophical framework: Husserl, Heidegger, and Merleau-Ponty.[1] In this chapter and those that follow, I show how the philosophy of Deleuze and Guattari resonates with the 4EA approach. Here I begin that task by way of linking them with Bruce Wexler's (2006) *Brain and Culture*. After a brief introduction, we will treat the following topics: (1) neuroplasticity, (2) "difference and development," and (3) political affect. The thesis of the chapter is that Wexler shares with Deleuze a "dynamic interactionism" such that human being is to be found in the processes of forming and reforming relations via individuation processes that integrate a social-neuro-somatic multiplicity on evolutionary, developmental, and behavioral temporal scales.

The standard schools of thought in cognitive science use a computer model for the brain–mind connection: brains, like computers, are physical symbol systems, and minds are the software run on those computers. The difference is in the respective computer architectures. Computationalism sees cognition as the rule-bound manipulation of discrete symbols in a serial or Von Neumann architecture, which passes through a central processing unit. Connectionism, the second standard approach, is based on another computer model, but it has a different, allegedly more biologically realistic architecture: parallel distributed processing. In connectionism's "neural nets," learning is a change in network properties, that is, in the strength and number of connections.

The 4EA approach, though agreeing with connectionism that a model of the mind must have a realistic chance of being instantiated in brains like ours, differs from both computer models by defining cognition as the direction of action of an organism in its environment rather than as a kind of information processing. Hence one key difference is that the 4EA approach breaks with any unidirectional information processing model in which cognition is the middle slice in what Susan Hurley (1998, 21) called the "classical sandwich": sensory input–processing of representations–motor output. Similarly, the 4EA school rethinks the allegedly central role in cognition of "representation," in which a model of the world is built up inside the cognitive agent. Rather, adopting Rodney Brooks's slogan that "the world is its own best model," the 4EA thinkers will restrict representation to a few "offline" problems and see the vast majority of cognitive processes as real-time interactions of a distributed and differential system composed of brain, body, and world.[2]

Let us use Deleuze's ontology, as we laid it out in "Introduction I," to unpack that last formulation in which individuations integrate a differential field. Even the neural section of *neuro-somatic-environmental system* is a distributed neural architecture whereby neurodynamic processes are seen as the integration or resolution of distributed–differential neural systems. We can see the embodied and embedded nervous system as a preindividual virtual field: (1) a set of differential elements (reciprocally determined functions—in other words, neural function is networked: there is no such thing as the function of "a" neuron; some argue the same for higher-level cognitive processes, i.e., that they emerge from global brain activity and hence cannot be understood in isolation) (2) with differential relations (linked rates of change of firing patterns) (3) marked by singularities (critical points determining turning points between firing patterns). The dynamics of the system as it unrolls in time are intensive processes or impersonal individuations, as attractor layouts coalesce and disappear as singular thresholds are passed (Varela 1995). Learning, then, is the development of a repertoire of virtual firing patterns as they relate to bodily interactions with the world. Any one decision is an actualization, a selection from the virtual repertoire, that is, the coalescing of a singular firing pattern; this is modeled by the fall into a particular basin of attraction from the attractor layout proposed by system dynamics.

Neuroplasticity

We have just seen how individuation as integration of a differential field works on the behavioral level. Let us now look to Wexler's book to consider his treatment of developmental individuation and its condition in neuroplasticity. Wexler makes a series of interconnected points: our sociality and our brain structure–function have coevolved such that humans have evolved for a long period (through young adulthood) of socially mediated neuroplasticity (Wexler 2006, 16, 142). In fact, the most socially sensitive plastic parts of the human brain are precisely the ones whose proportions relative to other brain structures distinguish humans from other primates (e.g., frontal and parietal lobes, involved in decision making, impulse control, etc.) (31, 105). So we see our multiple temporal and compositional scales—phylogenetic evolution, ontogenetic development, and behavioral action—intersecting emergent structures in the suprasubjective social, the subsubjective somatic, and the adjunct-subjective registers.

However, this developmental neuroplasticity is relatively reduced in adulthood. In a formula, children need sensorimotor and social stimulation to form neuropsychological structures, whereas adults look to shape their world in accordance with previously formed structures. These processes are linked and mutually reinforcing in that different cultures have different characteristic sensorimotor and social interaction patterns (e.g., the comfort zone for conversations—how close people stand to each other while talking—varies across cultures). It is important to emphasize that both children and adults work in a system of structuring structures: input produces a structure that shapes further input; once activated, these structuring structures are reinforced because consonance of input and structure produces pleasure and dissonance produces pain (155). The "structuring structure" scheme is clear here in Wexler's summary of studies of prejudice: "First, since these internal structures select and value sensory input that is consistent with them, they create an exaggerated sense of agreement between the internal and external worlds. Second, since internal structures shape perceptual experience to be consistent with the structures themselves, they limit further alteration of brain structure by environmental input" (155). The difference is that adults—or at least some adults, those in favored social positions—can act on the world according to this principle (or at least subconsciously select input conforming to previous structures), whereas children's actions in this regard are largely limited to adapting to what is the case (182). However, owing to the complexity and mutability of current cultural conditions, what is the case for children often differs considerably from the experiential structures of adults (6, 142). This difference in neuroplasticity and experiential structures sheds light on generational conflict, bereavement, immigrant experience, and social conflict in multicultural situations (144).

Turning from this summary of his points, we see that Wexler's opening chapter, a primer on neuroscience, shows the resonance with Deleuze clearly. Wexler emphasizes that higher-level cognitive function (thinking as well as sensorimotor perception–action) is an emergent process that occurs as the result of the integration of firing patterns in multiple brain structures; it cannot be located in individual neurons (22). This neuronal emergence is deepened by Wexler to a position that falls in line with the 4EA school, which locates cognitive functions as those of an organism in its environment, not simply as the result of brain activity alone. The following passage clearly shows Wexler's neuroenvironmental emergentism: "The specific patterns of all the intricate connections among neurons that constitute these functional systems are determined by sensory stimulation and other aspects of environmentally induced neural activity" (22). Here it is clear that cognition is emergent not simply from multiple brain systems but from a differentiated system in which brain, body, and world are linked in interactive loops. Indeed, the whole thrust of Wexler's position serves to "minimize the boundary between the brain and its sensory environment, and establish a view of human beings as inextricably linked to their worlds by nearly incessant multimodal processing of sensory information" (9).

Wexler's view of humans as embedded in their (largely social) environments has profound philosophical implications, as it entails that we must see human individuation as the continuous forming and reforming of singular patterns of somatic and social interactions. We can see the connection with Deleuze if we reformulate his ontology as what we can call a dynamic interactionist ontology. From this perspective, it is the interaction of intensive individuation processes with social, neural, and somatic dimensions to their multiplicity that forms the contours of the virtual field of potentials for further indi-viduations. In a formula, we are individuated as singular patterns of social, neural, and somatic interaction, each actualization structured by and structuring its virtual field. The embodied and the embedded aspects of our being intersect—we are bodies whose capacities form in social interaction. "Singular patterns of social and somatic interac-tion" means that we are what we can do with others—our embodied capacities, which develop in the history of the social interactions we have had up to the present, intersect with the similarly constituted embodied capacities of the others we now encounter. The creative potential of these encounters is such that we do not know who we are, we do not know what human nature is, until we experiment with what we can do with others. This is not to endorse a naive and outmoded social constructivism; but if evolved plasticity is our nature, as Wexler argues, then that nature is certainly more open than programmed, or perhaps better, it is programmed to be open to the construction of singular complex patterns as well as to the installation of triggers and thresholds for basic emotional patterns and for proto-empathic identification, as we explored in chapters 2 and 3.

Let us continue. Wexler's treatment of human studies in his chap-ter on the effects of the social environment on experience-mediated neural structures stresses the mother–infant dyad (2, 96–100). He is careful, and we have to be careful too, to avoid any "fusion" images, which distort individuation by seeing it as separation from a prior fu-sion, with all the anxiety about engulfment that entails. The dyad is a patterned interactive process, with the caregiver providing a scaffold, a supplement, which provides structure just beyond what the infant is capable of at any one moment and within which the infant can develop its capacities for self-regulation (109). In this way, self-regulation shifts from the parent-regulated dyad to the developing infant (105). Wexler uses the term *internalization,* but we must recognize that this is not the internalization of external structure by a formless infant; rather, the infant has some capacity for self-identification, some crude and vague but active body schema, or else it could not *relate* to others. In other words, if you are going to do away with isolated independence as the telos of human individuation, you should do away with infant–caregiver fusion as its arche.

The notion of an infant body schema refers to the Meltzoff–Moore experiments on neonate imitation, which Shaun Gallagher (2005) interprets in terms of an early body schema of the infant; these studies would belie any talk of "unity" and "oneness" between infant and caregiver. We should note the difference between the mere possession of an infant body schema and more mature subjectification processes. But there are some infant researchers, Colin Trevarthen, in particular, who insist on a subjectivity of the infant, its active participation in the give and take with a caregiver (Trevarthen 1999). Others, such as Horst Hendriks-Jansen (not a psychologist but a philosopher), read the same infant developmental literature and emphasize the caregiver's role in "scaffolding" the infant into subjectivity (Hendriks-Jansen 1996, 252–77). Daniel Stern's (1985) work on "the interpersonal world of the infant" is part of the picture here as well. All stress the way this sort of identification and projection of subjectivity via the face is well placed in forming an emotional bond and in beginning a "scaffolding" subjectivity-inducing loop between infant and caregiver. The important thing to keep in mind is that the structure the caregivers provide is a pattern of interaction; the infant does not internalize the properties of an independent substance. We come back to this formula: we are patterns of social–somatic interaction. Wexler (2006, 39) writes in a key passage, "The relationship between the individual and the environment is so extensive that it almost overstates the distinction between the two to speak of a relation at all." "Almost" is the key here; you must have relatively independent relata to have a relation, but you cannot have pure independence either.

We are now in a position to explore Wexler's version of dynamic interactionism. At this point, I want to nuance the formulation in the original version of this text (Protevi 2010), *radical relationality,* in which the relata are only nodes of multiple relations. This is close to what I have in mind, but the infant's body schema gives a certain singularity to it as a relatum, so it is not quite relations all the way down. The infant's singularity is not much, and it is certainly ready for further individuation via forming and breaking relations with others, but it is not quite simply a node of a relational network. Stating it that way collapses the difference between intensity and virtuality, between the process of differenciation and the state of being differentiated; it conflates the "order of reasons" we saw in "Introduction I":

"differentiation–individuation–dramatization–differenciation (organic and specific)" (Deleuze 1994, 251; 1968, 323; translation modified to restore the original word order of parentheses). The difference is this: as an intensive individuation–dramatization process, the infant has an irreducibly singular character; it is the ongoing actualization and hence differenciation of a purely differentiated virtual multiplicity with interlocking social, neural, and somatic dimensions. As Deleuze (1994, 252) puts it in *Difference and Repetition,* affirming a formulation of Lucretius regarding the singularity of fields of individuation, "no two eggs or grains of wheat are identical." The virtual register is a field of radical relationality; the intensive register is composed of irreducibly singular processes.

Over time, we are the patterns of the processes of forming and reforming of relations between individuals who are each ongoing and singular individuation processes. In light of his dynamic interactionism, Wexler (2006, 39) will question our fantasies of complete independence: "individuals often have an exaggerated sense of the independence of their thought processes from environmental input." This exaggerated sense of independence happens because of a substantializing illusion: "Thus, as we develop into unique individuals as a result of both our unique cumulative interactions with the environment and our unique hereditary characteristics, our uniqueness seems a property of us" (40). But that "seeming" to possess "properties" is an illusion, or better, the result of conflating stages of our ontology: we are never, no matter how habit ridden, fully differenciated static entities with unchanging properties but are instead ongoing singular processes of forming patterns of interaction, singular ways of integrating and differentiating a multidimensional differential social–neural–somatic field. So Deleuzian and Wexlerian dynamic interactionism does not deny individuality but sees it as an ongoing, self-modulating process of individuation–dramatization that is always in touch with a "metastable" field or "egg" that is itself the incarnation of a virtual multiplicity.[3]

Wexler's dynamic interactionism comes out in his refreshingly open-minded discussion of psychoanalysis, in which he distinguishes an individualist and "drive"-oriented theory from an interactional one (123). The process of development via structuring structure is one of the main modes of our relationality. What must always be kept in mind to appreciate Wexler's dynamic interactionism is that the target of "internalization" is never the properties of a separate substance but the "interpersonal or even multiperson processes that had *not previously existed in any particular individual*. That is, the qualities of the developing individual arise from interactive combinations of processes based on several individuals" (125–26; emphasis added). Dynamic interactionism thus entails emergent interindividuality, a notion we will see again in the discussion of political affect. The important thing to focus on here in connecting the dynamic interactionism of Wexler to that of Deleuze is the notion of individuation as the process of singularizing a pattern of interactional processes. Individuation is the creation of a pattern by which one navigates, by differentiation and integration, a multidimensional differential social–neural–somatic field. As Wexler puts it in another of his dynamic interactionist passages, "what was first external and interpersonal becomes internal structure. Adolescence and young adulthood are occupied with the dual tasks of integrating internal structures derived from multiple sources into a functionally coherent whole and articulating a personal ideology that leads to a niche in the general social matrix that is consistent with the internal structures" (136–37). To repeat the key point, the goal of development is to incorporate a novel pattern of interaction, "multiperson processes that had not previously existed in any particular individual" (125). Wexler puts it like this: human development is a matter of "the early shaping of the infant and child's psyche by the human-influenced environment, with the unique mixing of qualities from different adults and the internalization of historically influenced interpersonal processes" (128). The crux of Wexler's book is this creative novelty produced by dynamic interactionism, which, in its synchronic emergence, allows cultural variability and, in its diachronic, dynamic aspect, sets up generational interchange.

Difference and Development

Cognitive science, even that of thinkers allied with the 4EA approach, is still beholden to two unexamined presuppositions: first, that the unit of analysis is an abstract subject, "the" subject, one that is supposedly not marked in its development by social practices, such as gendering, that influence affective cognition, and second, that culture is a repository of positive, problem-solving aids that enable "the" subject. So we can use Deleuze to expand and deepen the 4EA school by turning to population thinking to describe the development and distribution of cognitively and affectively important traits in a population as a remedy to this abstract adult subject (chapter 6).

To make the connection with Wexler (2006), we note that chapter 4, "Self-Preservation and the Difficulty of Change in Adulthood," is the turning point of his book. Wexler first discussed socially mediated neuroplasticity in infancy, childhood, and adolescence so that individuation is a process of singularizing a multiperson pattern of social interaction. He now turns to discussing adult processes that seek to conserve mature patterns by selective attention to, or active shaping of, the world. The first adult process alters the perception of the existing world; that is, it works on the present from the perspective of the past. The second changes the world to "increase the likelihood that subsequent events will be consistent with pre-existing internal structures"; that is, it works on the present to make the future conform to the past (143). Wexler puts the transition from childhood to adulthood as a change in the relation of learning and power of action: "We learn the most when we are unable to act. By the time we are able to act on the world, our ability to learn has dramatically decreased" (143).

Wexler's take on our genetic endowment as the capacity for experientially mediated neuroplasticity resonates well with a school of biological thought, developmental systems theory (DST), that also resonates with Deleuze. The DST thinkers demand that we think about the social environment in which capacities develop (Oyama 2000; Oyama, Griffiths, and Gray 2001). Most of our capacities are not genetically determined; genes are a developmental resource, but there are other resources, intraorganismic and extrasomatic (e.g., recurrent social practices), that need to be taken into account (chapter 10). And once we are in the social realm, the cat is out of the bag. There can no longer be an abstract subject but rather there are populations of subjects, with varying distributions of capacities. And the practices that produce these capacities can be analyzed with political categories. Following Deleuze's biopolitical orientation, we can call the socially embedded person the *body politic*.

From this perspective of "difference and development" (see chapter 6), Wexler's (2006) bio-neuro-cultural standpoint would be helped by more emphasis on population variability, in two ways. First, the neuropsychological conservatism Wexler notes in adulthood varies within a population so that some adults remain in search of novel experiences. Now, as Wexler notes, sometimes this novelty is just a variation on familiar themes (17). But can we design a culture such that what people are used to is the search for novelty? I admit that you cannot just value novelty for its own sake. You do have to have familiarity and repetition, if only as repose from novelty searching. Furthermore, some novel experiences simply should not be experimented with (Protevi 2009, chapter 6). So we do need some normative standard: we should search for novel ways of empowering people to search for novel means of empowering others. In other words, our challenge is to make empowerment a radiating, horizontal social process. It is not like we are going to run out of such challenges in this quest; there is more than enough injustice to fight—we can let the ones who reach utopia worry about being bored. Thus we can say that some adults seek to "conserve" their inner neuropsychological structure by selecting friends who fight with them against unjust social structures and for positive social change. That is, they "conserve," in Wexler's sense, the fight against "conservatism" in the political sense. So what would be pleasant for them is not the conservation of an (unjust) social structure but the change of that social structure, to which end they seek to conserve the fight against that structure.

Second, attention to population variability is needed to attend to disempowerment right here at home. It is not just immigrants who face disempowering dissonance between internal structures and the external world. In any one society, many native people occupy "subject positions" that are devalued by the larger culture. Even though it is a great advance to talk about socially mediated neuroplasticity and the attendant notion of human ontology as the establishment of patterns of social interaction, we have to talk about populations of subjects, many of whom suffer disempowering subjectification practices. The key here is to propose a level of analysis that would not be merely idiosyncratic but that produces traits that would be reliably repeated and that would be open to political analysis. This is, of course, the major problem of feminism, race theory, queer theory, and other such analyses: where to locate the analysis so that you avoid the Scylla of personal anecdote and the Charybdis of ignoring difference altogether. Can we isolate structured subjectification practices that reliably reproduce populations with gendered and racialized subjects? We will pursue these questions in greater depth in chapter 6.

Political Affect

We can now turn to Wexler's take on political affect. As we saw in chapter 3, for Deleuze and Guattari, affect is desubjectified emotion. Wexler agrees. Let us follow the thread of dynamic interactionism we have seen in Wexler (2006) in his treatment of emotion as "an interindividual process that alters the moment-to-moment functional organization and activation patterns of the brain in the individuals who are interacting" (36). To appreciate the full radicality of this notion of emotion as an "interindividual process," we must add that those neural changes have to be thought in relation to the modifications to the emergent functional unit of the couple or group in which the component individuals are interacting. The neural bases of this interindividual process are found in each person's brain, but the unit we are analyzing is nonsubjective but relational, that is, interindividual.[4] To emphasize the interindividuality of such nonsubjective emotion, Deleuze and Guattari call it *affect*. We should also note at the outset that this emergent neuro-somatic-social emotional process need not only be equilibrium seeking; too often, any mention of group processes is seen as equilibrium seeking (negative feedback) as in "functionalist" sociology. Rather, we are all familiar with interpersonal emotions that spin out of control in positive feedback loops (a mob rage, of course, but on the positive side of the ledger, falling in love cannot really be seen as equilibrium seeking, even if a stable, loving couple results, for that stability can be a mutually reinforcing dynamic process of empowerment that never settles down to anything we can describe as an equilibrium).

In a later discussion of "a form of internally driven prejudgment of what is perceived" (Wexler 2006, 152), we once again see Wexler's dynamic interactionism. Wexler cites studies of "emotional valence" assigned to words and images in cross-racial contexts and concludes that "such prejudice demonstrates the effect of interpersonal modeling and sociocultural education on internal processes that alter the perception and valuation of stimuli" (154). We must be careful not to limit the notion of political affect to that of prejudice because, as Wexler notes, prejudice is an example of a "general process," in which "beliefs derive directly from sociocultural input, including the internal structures of important adults to whom the individual is exposed during childhood" (152). But as we saw earlier, these adults are not fully independent substances but are dynamically interactive in their turn. Adult structures, that is, adult patterns of interaction, are themselves individuations of a distributed and differential social field. The key point bears repeating. We do not internalize the structures of a single, substantial other. Rather, as we have seen, the target of internalization in Wexler's model is "interpersonal or even multiperson processes that had not previously existed in any particular individual. That is, the qualities of the developing individual arise from interactive combinations of processes based on several individuals" (125–26). Thus Wexler's dynamic interactionism implies that we are the patterns of the processes of forming and reforming relations via individuation processes that integrate a social–neural–somatic multiplicity.

⇒ 5 ⇐

The Political Economy
of Consciousness

The phrase "the political economy of consciousness" has a dual sense. It means both that the consciousness of individual actors plays a variable role in the "economy" of politics, that is, the analysis of factors that make up political activities, and that the production of the large-scale patterns of individual consciousness can often be analyzed in terms of subjectification practices that are tied to political economy. I will discuss the latter sense in the next chapter as the *granularity problem*. Here I look to political situations in which the effects of consciousness are attenuated or rendered superfluous in the economy of political action. I will provide three disparate and fairly self-contained analyses of such situations in this chapter: (1) discipline and rational choice theory; (2) a case study in the *socially invaded mind*; and (3) affect in Occupy Wall Street (OWS).

Discipline and Rational Choice Theory

Alva Noë (2009, 27–28) notes in *Out of Our Heads* the possible treatment of dogs as "a merely mechanistic locus of conditioned response." He goes on to say that we can do the same with human beings, noting that part of our horror at the Nazis lies in their "objectified, mechanistic attitude to human beings" (28). But we do not have to go that far. We can look at other much more mundane areas of sociopolitical practice that try to render irrelevant the effects of subjective agency by rendering behavior predictable. This black-boxing of consciousness can occur either en masse, by neoliberal economic practices that seek to produce the conditions that will in turn produce "rational," that is, predictable, behavior (for such an externalist reading of rational choice theory, see Satz and Ferejohn 1994), or in individuals and small groups, by discipline (Schwartz, Schuldenfrei, and Lacey 1979).

The idea is this: in certain forms of political activity, consciousness is not eliminated but is rendered superfluous in prediction and manipulation. In certain conditions, it simply does not matter what one would "prefer" in some private interiority because social constraints can be made strong enough to render the vast majority of actors predictable. (Bartleby's withdrawal did not change the productivity of Wall Street scriveners.) We see this in disciplinary institutions at the individual scale, for after a certain amount of training, most of the soldiers snap to attention, whether they like it or not. But it is not just the military; Schwartz, Schuldenfrei, and Lacey (1979, 229) investigate the nexus of the behavioristic emptying out of subjectivity and factory discipline:

> While behavior in the workplace now seems to conform to oper-
> ant principles, it did not in an earlier time, prior to the develop-
> ment of industrial capitalism. . . . The fit between operant theory
> and modern work is so close in part because operant principles, in
> the form of the scientific management movement, made modern
> work what it is. . . . Successful applications of operant theory do
> not necessarily confirm the theory. Rather, applications of operant
> principles to social institutions may transform those institutions
> so that they conform to operant principles.

On the social scale, consider Satz and Ferejohn's (1994) externalist reading of rational choice theory, in which, using an analogy with statistical dynamics, they show that in normalized conditions, the structure of a social system is all that need be analyzed. They dispense with the assumption of internal, psychological, rational agents; what they say needs to be studied are social conditions that produce behavior that can be modeled on the assumption of rational agents. "We believe that rational-choice explanations are most plausible in settings in which individual action is severely constrained, and thus where the theory gets its explanatory power from structure-generated interests and not from individual psychology" (Satz and Ferejohn 1994, 72).

Elinor Ostrom, Samuel Bowles, Herbert Gintis, and others in behavioral economics also have things to say to us here. A short piece by Ostrom (2005), "Policies That Crowd Out Reciprocity and Collective Action," has some important points relevant to our notion of the political economy of consciousness. Ostrom begins by reviewing evidence for strong reciprocators, the presence of which contradicts the rational choice theory assumption that rational egoists (utility maximizers driven only by external rewards and punishments) are the only type of agent that needs to be modeled to account for social behavior. Thus Ostrom proposes that we need to model different ratios of strong reciprocators and rational egoists and how those ratios change over time given different conditions. Strong reciprocators are conditional altruistic cooperators and conditional altruistic punishers. They are concerned with fairness of process rather than only with outcomes; in a word, they have internal motivations.

Ostrom continues: if you assume only rational egoists, then you have to design policies with external rewards and punishments. "Leviathan is alive and well in our policy textbooks. The state is viewed as a substitute for the shortcomings of individual behavior and the presumed failure of community" (Ostrom 2005, 254). The kicker is that such policies actually hurt the prosocial behaviors that would exist in their absence: "External interventions crowd out intrinsic motivation if the individuals affected perceive them to be controlling" (260). But internally motivated prosocial behaviors are not supposed to exist in a world of only rational egoists. So we have a self-fulfilling prophecy, or another example of "methodology become metaphysics": policies of externally compelled cooperation recommended on the assumption that social reality is a collection of rational egoists produce the very emptied-out, desubjectified reality that you have assumed. At this point, we should remember Satz and Ferejohn's externalism: what you study with rational choice theory is social constraint conditions. Properly set up, you can dispense with psychological attribution. To use a term of art in philosophy of mind, rational choice theory is the study of political economy zombies.

But all is not lost, Ostrom notes. If you design them properly, you can use external systems to "'crowd in' behaviors based on intrinsic preferences and enhance what could have been achieved without these incentives" (254). In other words, there really is, literally, a political economy of consciousness; with enough control, you can produce a combination of scarcity and disciplinary coercion that so constrains action as to render modeling of conscious decisions superfluous to prediction and control of behavior. In these situations, behaviorist manipulation via external rewards and punishments is not only suffi-cient for modeling predictable behavior but also crowds out reciprocity and collective action. Externalism can defeat internalism, if you will. Conversely, you can create institutional structures that provide the conditions for the survival and flourishing of internal motivations and concern for fair processes. In other words, you can—or better, you must—create the conditions in which conscious internal motivation can play an effective role in political economy.

A Case Study of the Socially Invaded Mind

U.S. representative Gabrielle Giffords was shot in an apparent assas-sination attempt on Saturday, January 8, 2011, in Tucson, Arizona. She survived, though six others were killed. After my initial horror at the case—a feeling that, my God, fascism is really here now, they are starting to assassinate their enemies—a short post at one of the blogs I frequent (Lawyers, Guns, and Money) piqued my philosophical interest. I commented there and then made some posts that elicited other responses on my own blog (New APPS) and as a guest on another (The Contemporary Condition). I was thus caught up in a give-and-take that began with issues of causality and eventually led me to the notion of the socially invaded mind, which I initially liked quite a bit but have subsequently come to question.

Although I will not spend too much time on it, there is a meta-level discussion to be had here in terms of the socially embedded mind: the process by which give-and-take on blogs helped my thoughts crystal-lize. They are my thoughts, but I would not have had them without this discussion. Or in other terms, there was social extension, me reaching out and making others think, and social invasion, thoughts bubbling up within me that were triggered by interactions with others.[1] I am going to present most of this in my own voice, as if the dialectic of proposal and objection was mastered by me all along, as if there were no extension and invasion, just a self-contained dialogue (the original version [Protevi 2011] preserves some of the give-and-take). But this sort of masterly presentation is a trick that hides the intersubjective process behind a seemingly self-contained product. So here we have yet another sense of the political economy of consciousness, con-sciousness as the fetishization of social labor, if you can accept the analogy with Marx's critique of the fetishization of commodities.[2] The academic practice of long lists of acknowledgments—to say nothing of long lists of notes—marks our anxiety about this troubling way in which we appropriate the intellectual commons, which makes it possible for us to be scholars, as private intellectual property, as "my ideas" (as if on cue, here is my reference: Read [2010]).

In any case, in the blog discussion of the Giffords case, I noticed a binary being produced: either we can show a direct ideological link between right-wing rhetoric and the (journals or video) expressions of the alleged shooter *or* the case is utterly mysterious and "senseless." For example, one social scientist said,

> To prove that vitriol causes any particular act of violence, we cannot speak about "atmosphere." We need to be able to demonstrate that vitriolic messages were actually heard and believed by the perpetrators of violence. That is a far harder thing to do. But absent such evidence, we are merely waving our hands at causation and preferring instead to treat the mere existence of vitriol and the mere existence of violence as implying some relationship between the two. (Sides 2011)

But this binary between hand waving and billiard-ball causality is a terribly impoverished view of causation. Biological thought helps us much more than this sort of physics model. Schmalhausen's law shows that we can make sense of the interchange of environment and population without meeting an impossible billiard-ball causality standard. (I am relying on the presentation of this concept in Lewontin and Levins [2007].) Schmalhausen showed that in species-typical environments, developmental robustness hides a lot of genetic variation. In other words, in normal environments, you can get roughly the same results in a population with genetic variance. But put that population under environmental stress, and the previously hidden genetic variation shows up in a greater range of phenotypes. This is not hand waving, but neither does it adhere to an impossible physics standard. The analogy is that the political rhetoric environment of Tucson was so extreme that we can plausibly suppose that it exposed the psychological variation in the population that would have otherwise remained unexpressed. Such an argument is not hand waving, and it should not be dismissed because it does not match some inflated standard of a direct cause-and-effect relation of one statement to one act. We could say that billiard-ball causality is "extensive" in its reliance on already formed objects and extensive properties of spatial and temporal location, whereas the sort of biological causality exemplified in Schmalhausen's law is "intensive" as it looks to triggering events that modulate ongoing intensive developmental and behavioral processes.

Now the psychological variation at stake concerns thresholds for violent action, which are very high in most people (chapter 2; see also Protevi 2009, chapter 6). In the overwhelming majority of people, only direct immediate physical threats provoke violence in return (and then not always): we are an extremely peaceful species when raised in moderately secure environments (Hrdy 2009; Fry 2007; 2012). But Representative Giffords posed no direct immediate physical threat to the alleged shooter, Jared Loughner, so we are looking for an indirect link, a matter of "influence." But where should we locate the link? Not at the level of ideology, I would argue (see the OWS section concluding this chapter). The link seems to be immersion in the antigovernment (and violence as the solution to the government-as-problem) milieu of Tucson. But we should not look for ideological motivation, as in a match between message intake and output, such as a repeated key phrase or even a possibly transformed key idea. Loughner did not have a coherent ideology. Nonetheless, he chose a Democratic politician targeted by right-wing rhetoric, and intensely so targeted by Giffords's opponent in the last election. So I think we have to look not to a smoking gun ideological match but to the way the target provided a promise to at least make a mark, to show he was serious, and so on. Any big target would do, but this one had a particularly salient energy attached to her. So, we could say, the ideology does not belong to Loughner, but he picked up on the energy that a particular ideology aimed at Giffords. It is not the ideology that counted to Loughner but the social energy that became attached to Giffords. And that energy was not generalized antigovernment sentiment but was specifically targeted by those who do have an ideological grudge against Democrats.

To come back to our leading question, why is billiard-ball causality so problematic in this case? Because it produces much too crude a view of political psychology, especially with regard to the role of belief in Loughner's reaction to the environment. Note the key claim of the initial blog post cited earlier: "We need to be able to demonstrate that vitriolic messages were actually heard and believed by the perpetrators of violence" (Sides 2011). Answering the question of whether vitriolic messages were "heard" by Loughner is quite easy. The shooter was described as "obsessed" with Giffords; he attended one of her rallies in 2007; and she won her election by a small margin, thirty-five hundred votes, against a candidate whose campaign had all sorts of violent images. It is a vanishingly small probability that Loughner was not exposed at some point to these sorts of things. Now as to Sides's second requirement, belief, here we are a lot closer to the billiard-ball causality I mocked previously. With the requirement that we prove that Loughner "believed" vitriolic messages, we are first called on to prove that he had a mental representation with the semantic content "Giffords must be eliminated." And we are then called on to trace the genesis of that representation to an event at time T1, the exposure to a particular message or set of messages. We would then have to show that this representation with that content, plus some other representations, are then the necessary and sufficient conditions for his action.

Now it may be that Sides has a more sophisticated psychology than the preceding sketch, though it is hard to tell from that blog post. And it is certainly no good on my part just to chant "nonlinear dynamics" as a mantra so that anything goes in linking environment and shooter. But there has to be something along the lines of developed dispositions and thresholds that is better for thinking this case than the sort of linear belief–desire–action scheme Sides seems to be proposing and that Susan Hurley (1998) memorably mocked as part of the "classical sandwich" view: sensory input—computation on representations—motor output. The important thing to remember is that the Giffords case is not an isolated incident; right-wing violence in the United States is a well-established phenomenon.[3]

What view of causality must we develop to discuss the singularities in this pattern, such as Loughner? I would say that the poisonous rhetoric here is a factor in a complex system. What I object to is the exclusive binary by which, unless one can show a strict linear causality, one can say nothing. I would be happy if people would say that there are sometimes linear causation systems (with some ceteris paribus conditions) but that they are a minority even in physics; the general case is complex nonlinearity. But we have then to expand our notions of causality rather than restrict them to linear causality versus mere correlationist hand waving. Now, with regard to the biology analogy, I do not think the unexpressed genetic variation gets all the credit here. In brief, it is the interchange between the environment and the genetic variation that is responsible, over development, for psychological variation with regard to violence thresholds. And that interchange is the individuation process of an eco-devo-evo multiplicity (chapter 10). That is a long way from just "genetic variation," and besides, there is a way the environment constructs the expression, which, ex post facto, reveals what had been unexpressed. This is certainly paradoxical on a linear causality model, but I argue that that is what we have to say in chapter 10. Now I am not calling, necessarily, for restrictions on political discourse and images, but I am saying we need to think about it, as Susan Hurley (2004) did when we called for thinking about the legal status of first-person shooter games. Finally, then, I would say that there is no sophisticated causality in which certain messages are *the* cause of Loughner's action—because that is not a sophisticated causality. Those messages are, arguably, causal contributing factors in a political affect multiplicity.

To conclude this section, let us examine the notion of the socially invaded mind. Continuing to bang away at this critique of the binary between having to show a direct link between specific pieces of rhetoric and Loughner's act versus having to content ourselves with general correlations, I thought I could adapt Susan Bordo's (1986; 1993) phrase, "psychopathology as crystallization of culture," which she used to resist the medicalization of anorexia. We would never be able to identify one image and the onset of anorexia in a particular anorectic, but I would not want to say that there was no connection at all between cultural images of desirable thinness (plus, e.g., those of thinness as sign of willpower) and that particular anorectic. So the idea is that Loughner was not outside culture in being insane. On the contrary, he was too close to it; he had no filters or not strong enough filters. He did not have a socially extended mind; rather, he had a socially invaded mind—the outside just came pouring in.

But having laid out this model, my thoughts on Bordo were considerably sharpened by this comment by Hasana Sharp in personal communication:

> My worry about the Bordo-model is that it could imply that the problem with [people as] social mirrors is that they are not Cartesian enough—that the solution is better filters, better abilities to affirm or deny the validity of our sensuous representations. It does not have to imply that: it could mean we need better buffers. His social constellation did not provide any alternatives and exacerbated these cultural tendencies, whereas we are inserted in other constellations that make tea party rhetoric sound either (a) like rhetoric/posturing/playing a game and/or (b) insane. . . . We need to resist the Cartesian conclusion that we need individually cultivated critical faculties that are permanently set on skepticism, or else we are profoundly vulnerable to the deceptions of opinion and sensation (= culture). I do not think Bordo is wrong, only that there is still some Descartes lurking there, despite her magisterial critique of him as a pathological symptom.

So, I thought, I have to stress not just Loughner's low filters, which enabled him to be "socially invaded," but also Tucson as the invading element. The object of analysis is the individuation process, "Loughner-as-he-develops-in-Tucson." But even that might not be enough: as Sharp argues, the socially invaded mind idea is still too individualistic. It is not just that he had a socially invaded mind but that the society that invaded him, Tucson, provided him no buffers; it was all "guns are the solution to government," all the time. Having no filters in Ann Arbor, Michigan, might have kept him in a basement making YouTube videos, but having no filters in Tucson put him in that supermarket parking lot.

But then, having questioned the reversal of polarities and recognized that we are all socially extended *and* socially invaded, we have to look again at the "we" from our population variation perspective so that when it is a sick culture invading a population, it is still only the case that only a few will crack under the stress. But with this population perspective, especially when it comes to the production of embodied violence thresholds, is *mind* the right term, rather than *biosocial subject*? How much dynamically affective enaction, how much of an eco-politico-devo-evo multiplicity, can we build into our models before *mind* becomes an untenable term for what we are after? It is partially for that reason that I tend toward the formulation of *body politic* in *Political Affect* and elsewhere.

Semantic, Pragmatic, and Affective Enactment at Occupy Wall Street

The Occupy movement shows us how the semantic, pragmatic, and affective—meaning, action, and feeling—are intertwined in collective practices. The intertwining of the semantic and the pragmatic—what we say and what we accomplish in that saying—has been a topic of interest in the humanities and the critical social sciences for almost fifty years, since its thematization by Austin and its codification in speech act theory; widespread interest in affect has been more recent, but the interplay of its twin roots in Tompkins and Deleuze—producing a sort of evo-neuro-Spinozism—has been usefully explored in *The Affect Theory Reader* (Gregg and Seigworth 2010; for a mildly critical take on "affect theory," see Wetherell 2012). It is now time to bring speech act theory and affect theory together in understanding the role of political affect in the Occupy movement.

To do that, we will first need to do some housecleaning. The first thing that needs to go is the concept of ideology. Deleuze and Guattari (1987, 68) say in *A Thousand Plateaus*, "Ideology is a most execrable concept concealing all of the effectively operating social machines." I take that to mean that we have to thematize political affect to understand "effectively operating social machines." From this perspective, the real "German Ideology" is that ideas are where it's at, rather than affect. It is political affect that "makes men fight for their servitude as stubbornly as though it were their salvation" (Deleuze and Guattari 1984, 29).

Why won't *ideology* cut it? It does not work because it conceives of the problem in terms of "false consciousness," where that means "wrong ideas," and where ideas are individual and personal mental states whose semantic content has an existential posit as its core, with emotional content founded on that core, so that the same object could receive different emotional content if you were in a different mood.

Thus, to take up the great OWS poster, "Shit is fucked up and bullshit," the core act posits the existence of shit, and then we express our emotional state by predicating "fucked up and bullshit" of it, whereas we could have predicated "great and wonderful" if we were in a different mood. But that is "execrable" for Deleuze and Guattari because it is far too cognitivist and subjectivist. It is too cognitivist because it founds emotion on a core existence–positing act, and it is too subjectivist by taking emotion to be an "expression," something individual that is pushed outward, something centrifugal. For them, emotion is centripetal rather than centrifugal, or even better, emotion is for them the subjectivation, the crystallization, of affect. Now Deleuze and Guattari do have a coporeal–Spinozist notion of affect involved with the encounter of bodies, but they also have what we could call a *milieu* or an "environmental" sense of affect. Here affect is "in the air," something like the mood of a party, which is not the mere aggregate of the subjective states of the partygoers. In this sense, affect is not emergent from preexisting subjectivities; emotional subjectivities are crystallizations or residues of a collective affect.[4]

Having done away with "ideology" as an analytical concept, we can turn to a simple, powerful talk by Judith Butler at OWS (Butler 2011b) that uses the phrase "enacting the political." Butler's talk calls on the classic "very well then, we demand the impossible" trope and ends with the wonderful line, "We're standing here together, making democracy, enacting the phrase, 'We the People.'" A longer talk by Butler in Venice (Butler 2011a) discusses constituting political space while acknowledging the material precarity of bodies, developed alongside a critical analysis of Arendt's notion of a political "space of appearance." The overall aim is set forth when Butler states, "A different social ontology would have to start from the presumption that there is a shared condition of precarity that situates our political lives."

A brief excerpt from the beginning of Butler's Venice talk sets out some of the main lines of thought that would go toward this "different social ontology":

Assembly and speech reconfigure the materiality of public space, and produce, or reproduce, the public character of that material environment. And when crowds move outside the square, to the side street or the back alley, to the neighborhoods where streets are not yet paved, then something more happens. At such a moment, politics is no longer defined as the exclusive business of a public sphere distinct from a private one, but it crosses that line again and again, bringing attention to the way that politics is already in the home, or on the street, or in the neighborhood, or indeed in those virtual spaces that are unbound by the architecture of the public square. . . . But in the case of public assemblies, we see quite clearly not only that there is a struggle over what will be public space, but a struggle as well over those basic ways in which we are, as bodies, supported in the world—a struggle against disenfranchisement, effacement, and abandonment.

Butler's notion of a differential social ontology is obviously one with which I am highly sympathetic. However, I would say that the role of the body in social ontology need not be limited to shared precarity, as important as that is to emphasize to break down notions of individuals as disembodied bundles of rights. We can also think of the positive affective contribution of public assemblies. In this case, the city government of New York unwittingly helped OWS tap into the affective potential of collective bodies politic. I'm talking here about the human microphone, which works, quite literally, to amplify the constitution of political space by assembled bodies.

The human microphone thus offers an entry into examining political affect in the enacting of the phrase "We the People" at OWS. It shows us how direct democracy is enacted by producing an intermodal resonance among the semantic, pragmatic, and affective dimensions of collective action. It also shows how the production of contemporary neoliberal subjects (*homo economicus* as self-entrepreneur, as individual rational utility maximizer) is so successful and so pervasive as to be invisible. The city thought it was hurting OWS by banning bullhorns, when in fact they helped them immensely by allowing the affect produced by entrained voices, a collective potential they could not grasp.[5]

As we saw in chapters 1–3, I am fascinated by studies of human entrainment, such as McNeill (1995), which studies the political affect dimension of entrainment (the falling into the same rhythm) by collective bodily movement as in communal dance and military drill. The neuroscientist Scott Kelso (1995) has studied all sorts of small-scale examples of entrainment (toe tapping and so on) by using dynamic systems modeling. A famous macroexample of spontaneous entrainment is the Millennium Bridge episode, in which the unconscious synchronization of walkers produced a resonance effect on the bridge, causing a dangerous lateral sway (Newland, n.d.). The developmental psychologist Colwyn Trevarthen (1999) has studied mother–infant intercorporeal rhythms in terms of "primary intersubjectivity."

The upshot of this research is that humans fall into collective rhythms easily and that such collective rhythms produce an affective experience, a feeling of being together, an *eros* or *ecstasis,* if you want to use classical terms, the characteristic joy of being together felt in collective action (Ehrenreich 2007). So I wonder if the human microphone (Ristic 2011), an invention of the OWS assembly when New York City banned electric bullhorns, does not contribute a little to the joyful collective affect of OWS. (Needless to say, the prospect that the human microphone might aid in the production of such collective joy frightens the right-wing commenters [Dyer 2011].) It is not quite a choir, but it is a chorus, and so the bodies of the chanters (their chests, guts, throats, eardrums) would be vibrating at something close to the same frequency, something close to being in phase.

Now I'm not a reductionist; the semantic cannot be reduced to the corporeal; the message is not dissolved into the medium. What interests me is how, in the human microphone, the message (enact the phrase "We the People") is resonant with and amplified by the medium (collective rhythm). In her Venice talk, Butler (2011a) analyzes the Tahrir Square chant translated as "peacefully, peacefully" in these terms:

Secondly, when up against violent attack or extreme threats, many people chanted the word "silmiyya" which comes from the root verb (salima) which means to be safe and sound, unharmed, unimpaired, intact, safe, and secure; but also, to be unobjectionable, blameless, faultless; and yet also, to be certain, established, clearly proven. The term comes from the noun "silm" which means "peace" but also, interchangeably and significantly, "the religion of Islam." One variant of the term is "Hubb as-silm" which is Arabic for "pacifism." Most usually, the chanting of "Silmiyya" comes across as a gentle exhortation: "peaceful, peaceful." Although the revolution was for the most part non-violent, it was not necessarily led by a principled opposition to violence. Rather, the collective chant was a way of encouraging people to resist the mimetic pull of military aggression—and the aggression of the gangs—by keeping in mind the larger goal—radical democratic change. To be swept into a violent exchange of the moment was to lose the patience needed to realize the revolution. What interests me here is the chant, the way in which language worked not to incite an action, but to restrain one. A restraint in the name of an emerging community of equals whose primary way of doing politics would not be violence.

This is an insightful, eloquent analysis of the pragmatics and semantics of the chant. So it is not to undercut it that I call attention to the material dimension of the resonating bodies that accompany the semantic content and pragmatic implications of this chant. It is to point to the way in which an analysis of material rhythms reveals the political affect of joyous collectivity and the intermodal (semantic, pragmatic, affective) resonance such chanting produces.

Finally, let me end with a few words on political affect. Joy in entrained collective action is by no means a simple normative standard. There is fascist joy—the affect surging through the Nuremberg rallies, building on and provoking even more feeling, was joyous. If there is to be any normativity in political affect, it will have to be active joy rather than passive joy; active joy I understand as "empowerment," the ability to reenact the joyous encounter in novel situations, or to put it in semi-California-speak, the ability to turn other people on to their ability to turn still others on to their ability to enact active joyous collective action, on and on in a horizontally radiating network, or to use Deleuze and Guattari's term, a *rhizome*. Now political affect does not occur in a vacuum. It is not a matter of implanting a new feeling in any empty body; it is a matter of modulating an ongoing affective flow. So the joy of OWS has to convert a mood of shame. What counts in the "effectively operating social machine" demonizing welfare in the United States is the shame attached to receiving public aid without contributing to society with your tax dollars. It is shameful to have lost your job or your home; you're stupid, a loser to have been in a position to lose it, and you're a lazy, stupid loser if you have not found another one or if you never had one in the first place. You do not arrive at this American shame by aggregating individualized, subjectivized packets of shame; you get shamed subjects as the crystallization of the collective affect of shame in the American air.

And so you do not combat this shame by trying to change individual people's ideas, one by one, with information about unemployment trends; you combat it by showing your face, by embodying your lack of shame, by putting a face on unemployment or homelessness. You counteract the existing collective affect by creating a positive affect of joyful solidarity. Shame isolates (you hide your face); joyful solidarity comes from people coming together. It is joy released from the bondage of shame, to follow up on the Spinozist references. What is especially heartbreaking, then, about the We Are the 99% Tumblr site (2012), is that so many people still have some shame, as they only peek out from behind their messages. Hence the importance of the Occupy meetings; shared physical presence, showing your whole face: these create the positive affect, the shamelessly joyful solidarity needed to fully overcome shame. Fighting the residual shame, the half-faces of private pictures sent to a website: that is what makes the collective occupation of space so important—bodies together, faces revealed, joyously.[6]

So I am going to propose that a full enactment of direct democracy means producing a body politic whose semantic ("we are the people; we are equal, free, and deserving of respect in our precarity and solidarity"), pragmatic (the act of respecting and supporting each other the assembly performs), and affective (the joy felt in collective action) registers resonate in spiraling, intermodal feedback.

= 6 =

The Granularity Problem

This chapter follows up on the second sense of the concept of the "political economy of consciousness" posed at the beginning of chapter 5. Whereas in that chapter, I considered cases in which the consciousness of individual actors played little or no role in political actions, here I ask about the relation of the production of the large-scale patterns of consciousness (the development and triggering of affective cognitive traits) by subjectification practices that are themselves analyzable in terms of political economy. The question here is the "granularity problem": what is the level of specification of analysis 4EA cognitive science should adopt when investigating concrete cases of the "extended mind"? Who gets access to the cognitive training (how does this thing work?) and the affective encouragement (the nurturing of confidence) needed to work with various external add-ons (from the famous notebook of Otto to all sorts of social institutions)?

In Deleuzian terms, the granularity problem concerns the relation of differentiation and differenciation: let us say we want to know what types of people tend to develop the computer skills for working with a particular software package. The question is, how far up do you go along the differentiations of the social-neuro-somatic multiplicity without simply saying something about a generic human subject's capacity for meshing with extended mind resources? Conversely, how far down can you go along the path of differenciation—the ontogenetic and behavioral scale—yet avoid being bogged down in the idiosyncrasies of singular situations? To have a political analysis of the subjectification processes that track the distribution of affective cognitive traits in a population needed for meshing with a mind extension, you need the right granularity, neither an unmarked generality nor a swarm of idiosyncratic singulars. In other words, there is no politics where each person has her own differenciated category, her own life story and "point of view." Conversely, there is no politics either with the unmarked generic "subject" of too much cognitive science, which only limns the least differentiated dimensions of "humanity" and "technology." So where do we get the right granularity—between idiosyncrasy and unmarked generality—for discussions of the social patterns for the use of and success with extended mind resources? After a brief discussion of some points of agreement between my basic approach and that of Shaun Gallagher, I intensify some aspects of Gallagher's approach and then conclude with examining the granularity problem and its relation to the methodology of case studies.

Points of Agreement

Shaun Gallagher is one of the most well known figures in the 4EA approach to cognitive science, so I will use his views as an orientation point. I will begin by noting two of the many convergences between my approach and that of Gallagher in his paper for the Socially Extended Mind workshop held at the Free University of Berlin in March 2011 (Gallagher 2011). Gallagher's perspective has also been elaborated in great detail in a series of works on social interactionism and participatory sense-making by Hanne De Jaegher, in collaboration with Ezequiel Di Paolo, Tom Froese, and Gallagher himself (De Jaegher, Di Paolo, and Gallagher 2010; De Jaegher and Froese 2009; De Jaegher and Di Paolo 2007).

Two of Gallagher's points seem to me most salient here: first, his insistence on the enactive—or what we could call, picking up on the analyses of chapter 4, the "dynamic interactional"—character of mind, countering the somewhat static view of the classical extended mind hypothesis (EMH); and second, the move to a distributed notion of judgment, countering the lingering individualism of the classical EMH (for current specialist debates on the EMH, see Menary 2010). Thus I am in agreement with the way Gallagher criticizes the canonical formulation of the EMH (Clark and Chalmers 1998) for its belief–desire psychology and its reliance on the whole apparatus of propositional attitudes, representations, intentional states, and so on. The classical EMH, Gallagher says, assumes a model of the mind that it should instead be challenging. For Gallagher, minds are—or perhaps better, emerge in and as—dynamic interactions between organisms and environments. As he puts it, they are composed of "enactive cognitive

processes and activities, e.g., problem solving, interpreting, judging"
(Gallagher 2011, 3). Although Gallagher does not explicitly explore
these ontological issues, we could say that he sees in the classical EMH
a view of the mind as a series of static states that successively measure
the adequation between internal beliefs and perceived states-of-affairs,
then calculate the difference between the current state and a desired
state, followed by a final calculation of what action would be best to
bring about the coincidence of the world and the desires of the subject. Furthermore, the state versus process debate—what we could
call the ontology of time as it relates to that of mind—comes into
play here, for the EMH view of serial information processing, even
when extended to include extrasomatic sources of information (Otto's
notebook), would, to speak Bergsonian for a moment, reduce time
to a series of slices in which operations on representations would be
performed, rather than respecting its concrete durational quality, the
time of dynamic process seen from the inside (Bergson [1913] 2001).

Similarly, I like the way in which, in insisting on the dynamic problem-solving character of the mind, Gallagher uses the legal system as
an example of a "mental institution" (3–7; see also Crisafi and Gallagher 2009; Gallagher and Crisafi 2009) to combat a lingering individualism in the EMH, whereby what is paradigmatically at stake is just me
and my notebook rather than a socially distributed process. As with
the preceding point, there is room for further discussion on some of
the ontological issues at stake here, as there is an ongoing "transversal"
emergence (Protevi 2006; see also "Introduction II"). Now synchronic
emergence is part–whole emergence so that at any one time, our judgments are "enabled and shaped" (Gallagher 2011, 6) by the systems in
which we are parts; the legal system is Gallagher's example. We also
find diachronic emergence as the system grows over time, as novelties
emerge, and there is transverse emergence as what links the two—in
Gallagher's words, "a dynamic process involved in . . . dialectical, trans-
formative relations with the environment" (6), the environment here
being the legal system and the transformations being the establishment
of a new judgment that will serve as a precedent for a series of "normal
judgments." In Deleuzian terminology, *new judgment* qua precedent is
a singularity or turning point that governs a series of ordinary points
(Lefebvre 2008; Protevi 2009, chapter 5).

In concluding this section, I should also note that I agree with

Gallagher's objection to the functionalism of the classical EMH and his insistence that mind is "enactively generated in the *specific* interactions of organism–environment (where environment is social as well as physical)" and that mind is "an enactive and emotionally embedded engagement with the world" (Gallagher 2011, 8; emphasis added).

Intensifications

I would now like to shift gears and intensify some of the elements of the enactive social cognition approach led by Gallagher, De Jaegher, and others. Let us consider this formulation by Gallagher (2011, 8) in his Berlin conference paper: "the mind as an enactive and emotionally embedded engagement with the world through which *we* [emphasis added] solve problems, control behavior, understand, judge, explain, and generally *do* certain kinds of things." There are five aspects to be intensified, aspects that Gallagher and De Jaegher often pick up on elsewhere—(1) synchronic variation, (2) diachronic development, (3) political categorization, (4) negative impacts, and (5) the socially invaded mind:

1. *Synchronic variation.* We (philosophers) can break up the "we" (posited as the cognitive subject of the EMH) by population thinking, that is, by focusing on the variation in the performance of affective–cognitive traits in a population of subjects. That is, just how many people in the population under examination engage smoothly or haltingly or not at all with various modes of extension? This focus on distribution of traits can help intensify the critical engagement with the socially extended mind, countering the tendency to think a generic "subjectivity" or "the" subject.

2. *Diachronic development.* Here we want to trace the development of the capacities by which understanding, judging, and engagement with modes of extension occur. To keep with the first point's emphasis on variation, we need to focus on the embodied affective–cognitive development of a population of subjects in a field of multiple, overlapping, and resonating or clashing subjectification practices, countering the tendency to focus on adult subjectivity.

3. *Political categorization or the granularity problem.* It does no good

to go from an unmarked generic subject or the "we" to the sheer dispersal of individuals (or even more radically, subindividuals: the plethora of drives, tendencies, mechanisms, and what have you, that make up of the underworld of our subjectivity). Rather, we have to thematize politically important categories, such as race and gender, lying between generic human subjectivity and idiosyncratic personality or subpersonality.

4. *Possible negative consequences.* Yes, sometimes social extensions help some subjects "solve problems," but we also need to be aware of the way politically charged subjectification practices can limit access to, or produce negative reactions to, some socially extended affective–cognitive institutions.

5. *Reversal of direction.* Finally, we need to be able to think the limit case of a socially invaded mind, a reversal of polarity of the usual inside-out vector used in thinking the extended mind, in which the initiation of action comes from a subject reaching out for help with cognition to elements in the world. (I have discussed this aspect in chapter 5's account of the "political economy of consciousness.")

As we move through the rest of this section, I will provide an integrated discussion, instead of taking each point individually.

Gallagher and De Jaegher note the lingering individualism of almost all cognitive science, referencing Boden (2006), who also notes it. However, we shouldn't forget that the original proponents of the EMH did in fact recognize "distributed cognition" when Clark and Chalmers (1998, 10) referred to Edwin Hutchins's (1995) celebrated example of the cognitive cooperation of a ship's crew as "research on the cognitive properties of collectives of agents." As we recall, for Hutchins, sailors are embedded in a technosocial assemblage—some working with charts, others with boiler room gauges, and so on—which enables them to work with their comrades so that an intelligent guidance of the ship emerges from their coordinated but decentralized cooperation. Though this synchronic emergence from a decentralized network is an important concept to retain, the EMH, even in its distributed cognition guise, has not examined the political context for the diachronic emergence of each crew member's affective cognition skill set (the ability to remain cool under pressure while still getting pumped up in times of crisis, indeed, the simple ability to get along well with others, are just some of the affective traits necessary for successful distributed cognition).

Building on analyses first put forth in Protevi (2009), let us note that EMH thinkers strive for a model of mind that is biologically "plausible" (Clark 1997; 2003). Thus they have incorporated population thinking into the evolutionary register, a prime theme of Clark's (2003) *Natural-Born Cyborgs*. However, they have not yet carried this over into the developmental register, into the political context of the development of affective cognition capacities in a population of subjects. To understand fully the complex interplay of "brain, body, and environment," we have to understand the diachronic and not just synchronic social environment. That means we have to study populations of subjects and the way access to skills training and cultural resources is differentially regulated along political lines.

There is a biosociality that we will have to consider here. Developmental systems theory (DST) proponents remind us that we have to think about the social environment in which affective–cognitive capacities develop. They are not genetically determined; genes are a developmental resource, but there are other resources, intraorganismic and extrasomatic (e.g., recurrent social practices), that need to be taken into account. And once we are in the social realm with regard to development, the cat is out of the bag. There can no longer be an abstract subject, but we must deal with populations of subjects, with varying distributions of capacities. And the practices that produce these capacities can be analyzed with political categories.

Lacking a population perspective on the development of affective cognition capacities, the EMH impoverishes its notion of "cultural scaffolding" by relegating the cultural to a storehouse of heuristic aids for an abstract problem solver who just happens to be endowed with certain affective cognition capacities that enable it to interact successfully with the people and cultural resources to which it just happens to have access. Positing an abstract or generic subject neglects the way in which culture is the very process of the construction of biosocial subjects so that access to certain cultural resources and to the training necessary to acquire certain forms of affective cognitive capacities is distributed along lines analyzable by political categories. This is not simply technical training for cognitive capacities in a restricted sense but also the training necessary for acquiring positive and empowering emotional patterns, thresholds, and triggers.

As an example of the sort of abstract, apolitical subject posited by the standard formulations of the EMH, let me cite a passage from Clark's (2003) *Natural-Born Cyborgs*. A fascinating work brimming with insights, Clark's book thinks together DST biology, interactive technology, and cognitive heuristics embedded in cultural scaffolding but still doesn't think the politics of subject production, positing instead a homogeneous subject, "us humans." Clark writes,

> A more realistic vision depicts us humans as, by nature, products of a complex and heterogeneous developmental matrix in which culture, technology and biology are pretty well inextricably intermingled. It is a mistake to posit a biologically fixed "human nature" with a simple wrap-around of tools and culture; the tools and culture are indeed as much determiners of our nature as products of it. Ours are (by nature) unusually plastic and opportunistic brains whose biological proper functioning has always involved the recruitment and exploitation of nonbiological props and scaffolds. (86)

This is all true, as far as it goes. But it needs to go further to examine the politics that regulate access to those "nonbiological props and scaffolds" and thereby regulate subject production. And not every subjectification practice is empowering. That is to say, some cultural practices harm individuals, instilling affective cognitive traits that help keep them in subservient positions via an internalization of negative self-image, for example. The recent analyses of "stereotype threat" in Cordelia Fine's (2010) *Delusions of Gender* could also be brought to bear here.

Now it could be said that I am being too harsh and should merely propose the population study of the production of subjectivities as a next step rather than as the making up of a gap left in the study of cognition. After all, one could object, you cannot really blame people interested in cognitive science for looking for the abstract principles of cognition and leaving the empirical study of actually existing subjectivities to psychologists. Here we see a sort of replay of the critique the embodied–embedded mind school made of functionalists. For functionalists, the abstract principles of cognition were the important thing, and neurology need not be consulted, as it provided merely the implementation details of mind, the hardware to which cognitive software was indifferent. The classical proponents of the EMH are functionalists to the extent that they say that in some cases, our cognitive software runs on a set of elements that cross the somatic boundary or "skinbag." The enactivists, however, insist that biology is relevant to the study of mind so that whatever cognitive software is proposed cannot be indifferent to its "wetware" instantiation. Thus, as we noted earlier, Gallagher (2011, 8) has shown that the EMH holds to a functionalism to which he objects, insisting that mind is "enactively generated in the specific interactions of organism–environment (where environment is social as well as physical)."

To conclude this section by way of repeating one of its main points, I think we can take Gallagher's insistence on examining "specific interactions" in temporal terms. Many EMH people look to evolutionary time scales (evolution of brain plasticity) and to behavioral time scales (the individual cognitive processes of me and my notebook) but overlook the developmental time scale. In other words, it is not enough to say that during evolution, human brains became plastic enough so that today they can take part in extended cognition episodes on the behavioral time scale. We also have to examine the kinds of social practices we see in the development in a population of subjects of a distribution of capacities for extended cognition episodes. Here the question is the level of concretion required of philosophers, a question to which we turn in the next section.

Granularity

In the preface to *Out of Our Heads*, Alva Noë (2009, xii) criticizes neuroreductionistic treatments of depression:

> It is simply impossible to understand why people get depressed—
> or why this individual here and now is depressed—in neural terms

alone. Depression happens to living people with real life histories facing real life events, and it happens not only against the background of those individual histories but also against the background of the phylogenetic history of the species.

I agree with the critique of neuroreductionism but want to propose a mid-level analysis between sheer idiosyncratic life histories and the evolution of the species. What are we to do with the fact that U.S. women are diagnosed with depression at twice the rate of U.S. men?[1]

To address the social-neuro-somatic dimensions that make up the multiplicity in which gendered rates of depression are lines of differentiation, we need a mid-grained conceptual scheme, between the too-fine-grained idiosyncrasies of individuals and the too-coarse-grained discourse of "the human." The key here is to propose a level of analysis that would not be merely idiosyncratic but that would produce traits that would be reliably repeated (to use the distinction Paul Griffiths and Russell Gray [2001, 196] use in discussing DST in biology) and that would be open to political analysis. This is, of course, the major problem of feminism, race theory, queer theory, and other such analyses: where to locate the analysis so that you avoid the Scylla of personal anecdote and the Charybdis of ignoring difference altogether. Can we isolate structured subjectification practices that reliably reproduce what we can call a feminized or masculinized body subject? And can we propose that as a philosophical desideratum for the discussion of 4EA cognitive science?

Though we can leave individual personality idiosyncrasies to therapists, can philosophers really leave aside, for example, the gender differences produced by contemporary subjectification practices? If all concrete subjectivities are gendered—or at least have developed via gendering practices and have to navigate a world in which gender matters—can we be satisfied with abstract principles of cognition that ignore gender effects? It is not enough to criticize computationalism and connectionism in the name of a nongendered embodied subject. We must realize that bodies are gendered and that such gendering changes the sphere of bodily "competence" within which objects appear, as detailed in Iris Marion Young's (2005) critique of Merleau-Ponty, "Throwing Like a Girl" (see also Butler 1989). To recall the outlines of Young's critique, for Merleau-Ponty, it is the practical abilities of the embodied subject that allow objects to appear as correlates of that subject's possible actions. But those capacities have to develop, and gendered subjectification practices will affect how those capacities do—or do not—develop and how they are distributed in a population of subjects. Thus feminized and masculinized embodied subjects can have different "spheres of competence": a flat tire can appear as a mildly irritating challenge or as an insurmountable problem, a subway entrance as an enticing gateway to the city or as an anxiety-producing danger. But this foregoing treatment is still too simple. It does no good to replace a single generic human subject with two abstractions, "the" feminized and "the" masculinized subject. We need to think in terms of a range of gendering practices that are distributed in a society at various sites (family, school, church, media, playground, sports field) with variable goals, intensities, and efficacies. These multiply situated gendering practices resonate or clash with each other and with myriad other practices (racializing, classing, religionizing, nationalizing, neighborhoodizing ["that's the way we roll"]). Adopting our Deleuzian scheme, we have to think a complex virtual field of these differential practices; in dynamic systems terms, we could think a complex phase space for the production of biosocial subjects—or in my terminology, *bodies politic*—with shifting attractor layouts as the subjectification practices clash or resonate with each other. But even this is still too simple, as these gendering practices also enter into complex feedback relations with the singular body makeup of the people involved, as I argued above in chapter 4 in distinguishing fully differentiated multiplicities from singular individuation processes.

Although I have stressed the need for mid-level political categories in 4EA analysis, between too-finely-grained idiosyncratic personality and too-coarsely-grained evolutionary stories about "the" human, I nonetheless think countereffectuating case studies, moving from singular differenciations up to the differentiating multiplicities of which they are the result of individuation processes, can be useful additions to philosophical methods. They can be an alternative to thought experiments such as brain transplants, brains-in-a-vat, zombies, Swampman—that whole bestiary—as well as an alternative to X-phil experiments, as when brain scans are done of people tackling

the trolley problem. Case studies do not aim at identifying the necessary and sufficient conditions for an essential distinction, as do thought
experiments, nor are they satisfied with collating fixed responses to
ready-made problems (x percent of the subjects pulled the lever, and
when they did, such-and-such brain area fired [e.g., Greene and Haidt
2002]). Instead, case studies as countereffectuations reveal the outlines
of problematic fields, which are the points of intersection of multiplicities in which linked rates of change create conflicting pressures
so that (1) any one move changes the conditions for future moves
and (2) no one solution exhausts the potentials for future creatively
different solutions.

James Williams (2005) gives this example of a Deleuzian problem:
"should we raise interest rates"? Deleuzian problems, the problems of
life, cannot be solved once and for all; they can only be dealt with. The
granularity problem in political investigations of the EMH is one such
problem: we can never solve it once and for all, but we can descend
from the fully differentiated social-neuro-somatic multiplicity, and we
can ascend from actual differenciated idiosyncratic cases. The task is
to find a degree of focus that allows us to analyze how a distribution
of affective cognitive traits useful for meshing with extended mind
resources comes about. That is, we can find the points in the differentiation to differenciation process of a social-neuro-somatic multiplicity
that seems appropriate for the research question posed.

≈ 7 ≈

Adding Deleuze to the Mix

This is the final chapter of part II. Whereas the other chapters were quite informal, owing to their origins as talks or blog posts, this is a more formal work for a specialist 4EA cognitive science journal, *Phenomenology and the Cognitive Sciences*. I will suggest ways in which adding Deleuze's philosophy to the mix can complement and extend the 4EA use of these core resources. But *why* add Deleuze to the mix? Is it worth the trouble? There is no gainsaying the complexity of Deleuze's thought or the strangeness of his terminology, but I hope to show that the benefits of adding Deleuze to the mix outweigh the costs. I will use Deleuze's ontology, as sketched in "Introduction I," to examine three areas of concern to the 4EA approaches: (1) the Deleuzian concept of the virtual will clarify the ontological status of perceptual capacity as sensorimotor skill; (2) the Deleuzian critique of the confusion of the actual and the virtual will enable us to intervene in the realism–idealism debate;[1] and (3) the Deleuzian concept of *intensive individuation* will clarify the ontological status of the genesis of perceptually guided behavior.

The prime focus for this chapter will be the ontological difference between a capacity and the exercise of that capacity.[2] Briefly stated, Deleuze enables us to move beyond two standard concepts of capacity: that of a self-identical or fully individuated possible awaiting realization and that of a self-identical or fully individuated potential teleologically oriented to its actualization. In place of these concepts, Deleuze proposes that a fully differentiated virtual field does not resemble that which is creatively actualized from it via an intensive process of individuation. Thus an individuated perception does not resemble the distributed and differential brain–body–world system, when that is conceived at the level of a virtual web of linked rates of change of neural, somatic, and environmental processes. Keeping this in mind, we will show how confusing the virtual structure of an intensive process (perception) with the actual properties of products (the represented world and the representing subject) will maroon us in the sterile realism–idealism debate.

Deleuze and the 4EA Approach

Although there are many significant differences in the 4EA approach to cognitive science ("embodied–embedded–extended–enactive–affective"), there is also a family resemblance in philosophical and scientific resources. Among the most prominent common reference points are, in mathematical modeling, dynamic systems theory; in biology, an enactive stance, often coupled with a positive attitude toward developmental systems theory; in psychology, Gibsonian ecological psychology; and in philosophy, phenomenology.

We can see some core elements of this family resemblance in a programmatic statement early in Michael Wheeler's (2005) work, *Reconstructing the Cognitive World*. Wheeler names his project "embodied–embedded cognitive science," which he opposes to "orthodox cognitive science" on four points: (1) that online intelligence, composed of "a suite of fluid and flexible real-time adaptive responses to incoming sensory stimuli" (12), is primary with regard to offline intelligence, which, as "detached theoretical reflection" (142), is "representation-hungry" (213–14), as in Wheeler's examples: "wondering what the weather's like in Paris now or weighing the pros and cons of a move to another city" (12); (2) that online intelligence is "generated through complex causal interactions in an extended brain–body–environment system" (12); (3) that embodied–embedded cognitive science displays an "increased level of biological sensitivity" (13); and (4) that embodied–embedded cognitive science requires a dynamic systems perspective (13–14).

However, insofar as it is a family resemblance, each work in the 4EA field need not have all elements. Anthony Chemero's (2009) *Radical Embodied Cognitive Science,* for example, brings dynamic systems theory and Gibsonian ecological psychology closely together (28, 83) but has barely a word on phenomenology, whereas Hubert Dreyfus's works were marked for many years by explicit and sustained reference to Heidegger and Merleau-Ponty yet have only very recently included a discussion of Walter Freeman's dynamic system neuroscience (Dreyfus 2007). Nonetheless, in quite a few key works, dynamic systems theory, biology, and phenomenology are all present, albeit with different emphases in the latter field: Andy Clark's (1997) *Being There* and Alva Noë's (2004) *Action in Perception* refer favorably, although mostly in passing, to Merleau-Ponty, whereas Wheeler (2005) and Evan Thompson's (2007) *Mind in Life* are heavily based on Heidegger and Husserl, respectively. Deleuze's focus on concrete perceptual process as an intensive individuation is consonant with the standard 4EA critique of orthodox cognitive science. That critique holds that the latter falsely takes representational thought as basic rather than derived: "the historical mistake of orthodox cognitive science has been its enthusiasm for extending its distinctive models and principles beyond the borders of offline intelligence, and into the biologically prior realms of online intelligence" (Wheeler 2005, 247). Andy Clark (1997, 1–2) echoes the critique of theoretical intelligence as derived but illegitimately seeing itself as primary when he calls for "a new vision of the nature of biological cognition: a vision that puts explicit data storage and logical manipulation in its place as, at most, a secondary adjunct to the kinds of dynamics and complex response loops that couple real brains, bodies, and environments." These critiques follow the general lines of the classical phenomenological critique of purely rational–representational theorizing as abstracting from its concrete practical ground and breaking free to posit itself as self-sufficient so that it pretends to ground that which in fact grounds it (Heidegger 1996; Merleau-Ponty 1962). Now it is certainly true that many within the classical phenomenology camp would object to Wheeler's use of the term *biologically prior* to characterize the concrete level, but this objection would not bother holders of the "deep continuity" thesis of the mind in life position, which builds on Jonas, in particular, to extend "lived experience" to all organisms (Wheeler 1997; Thompson 2007, 15, 157–62).

The phenomenological drive to concretion finds its home in 4EA cognitive science as a critique of functionalism, which came under fire for proposing a goal of constructing a cognitive architecture so abstract that it could license the treatment of concrete biology as mere "implementation details." The 4EA drive to concretion requires, rather, that we look for "biologically plausible images of mind and cognition" (Clark 1997, xvii; see also Wheeler 2005, 13; Thompson 2007, 5).[3] Considered as a response to computationalism, the allegedly more biologically adept connectionist paradigm is also criticized as abstract; for example, Wheeler's insistence on the way neurotransmitters produce modulations of neural firing patterns affecting a "volume" of processing structure, as opposed to the point-to-point binary logic of artificial neural nets, enables him to criticize connectionism as abstracting from the relevant biological details (Wheeler 2005, 13, 262–63).

However, as we will see, it is precisely the move to the abstract ontology of dynamic systems that enables us to see the biological details that matter. What we will see is that individuated, perceptually guided sensorimotor behavior is the actualization of a virtual, differential potential. This move across the ontological difference of virtual and actual has the structure of an integration of differential relations (i.e., linked rates of change) among coordinated neuro-somatic-environmental networks.

Deleuze and Neurodynamics

Let us take up the use of dynamic systems methods in neurodynamics as an area relevant to 4EA concerns, one in which Deleuze's concepts can help us with the ontology involved (Protevi 2009). Neurodynamics shows the brain as generating coherent wave patterns out of a chaotic background. During any one living act (perception, imagination, memory, action), the brain functions via the "collapse of chaos," that is, the formation of a "resonant cell assembly" or coherent wave pattern (Varela et al. 2001). Walter Freeman (2000a; 2000b) offers a dynamic systems account of the neurological basis of intentional behavior, while Alicia Juarrero (1999) uses dynamic systems to intervene in philosophical debates about decisions and intentional action. The basic notion in their accounts is that nervous system activity is a dynamic system with massive internal feedback phenomena, thus constituting an "autonomous" and hence "sense-making" system, in Varela's terminology, *when* it is seen as embodied and embedded. That is, sense-making is the direction of action of an organism in its world; in organisms with brains, then, the object of study when it comes to sense-making is the brain–body–environment system (Thompson and Varela 2001; Chemero 2009). Sense-making proceeds along three lines: sensibility as openness to the environment, signification as valuing, and direction as orientation of action. The neurological correlates of sense-making show neural firing patterns, blending sensory input with internal system messages, as emerging from a chaotic background in which subliminal patterns "compete" with each other for dominance.[4] Once it has emerged victorious from this chaotic competition and established itself, what Varela (1995) calls a "resonant cell assembly" forms a determinate pattern of brain activity that can be modeled as a basin of attraction.

Following the Freeman line of neurodynamic thought, supplemented by the embodied–embedded perspective, we see that in navigating the world, a person continually forms intentions, that is, leans toward things in outreaching behavior, as the brain–body–world system settles into patterns. Once in a pattern, the system constrains the path of future behaviors, as long as the behavior guided by the resonant cell assembly lasts. (Some intentions entail long strings of firing patterns, yielding coherent complex behavior, as in the intention to play a game of basketball.) Sensory input (changes in body correlated with changes in the world) continually feeds into the system along the way, either reinforcing the settling into a pattern or shocking the brain out of a pattern into a chaotic zone in which other patterns strive to determine the behavior of the organism. The neurological correlate of a decision is precisely the brain's falling into one pattern or another, a falling that is modeled as the settling into a basin of attraction that will constrain neural firing in a pattern. There is no linear causal chain of input, processing, and output; instead, there is continual looping, as sensory information feeds into an ongoing dynamic system, altering or reinforcing pattern formation—in model terms, the trajectory of the system weaves its way in and out of a continually changing attractor landscape whose layout depends on the recent and remote past of the nervous system.

Continuing with the perspective of somatically and environmentally supplemented neurodynamics, and taking up the sketch of Deleuze's ontology we laid out in "Introduction I," we make the link with Deleuze by seeing the neuro-somatic-environmental system as a virtual multiplicity: (1) a set of differential elements (reciprocally determined functions; in other words, neural functions are networked, i.e., they emerge from global brain activity and hence cannot be understood in isolation—and neither can global brain activity be understood in isolation from its somatic and environmental relations); (2) with differential relations (linked rates of change of neural firing patterns as they mesh with rates of change in body, world, and body–world interaction); (3) marked by singularities (as critical points determining turning points between patterns of relations among brain, body, and world).[5] The dynamics of the system as it unrolls in time are intensive processes or impersonal individuations. That is to say, behavior patterns emerge at a singularity or threshold in the differential relations as they are incarnated in the linkages of rates of change of functional neurosomatic processes; this coalescing of a behavior pattern is modeled by the fall into a particular basin of attraction from the attractor layout "proposed" by system dynamics. Over time, the repetition of a number of such actualizations provides a virtually available response repertoire, a set of capacities, for the person. With regard to any one actualized behavior pattern, the repertoire is virtual, and any one decision is an actualization, a selection from the repertoire. But "virtually available" cannot mean that the behavior patterns are individuated before their triggering. To respect Deleuze's ontological difference, we must say that before their triggering, behavior patterns are nothing, that is, the multiplicity or virtual field from which they emerge is fully differentiated and hence does not have the same ontological status as an actual, differenciated pattern. Furthermore, owing to countereffectuation, we cannot say that the repertoire is fixed: the structure of the multiplicity, the attractor layout, changes as the result of the intensive individuation processes we call "experience."[6]

Perceptual Capacity

The notion of perceptual capacity as a response repertoire or set of capacities is a key point at which the addition of Deleuze's philosophy to the core 4EA resources pays off, for the ontological status of "virtual" is needed to understand the mode of being of capacities that are not actually at work. That is, we need to see perceptual ability as a multiplicity and perceptual action as an individuation of that multiplicity. In this way, Deleuze enables us to overcome two classical concepts that relate a capacity to its exercise: possibility awaiting realization and teleologically oriented potentiality.

First, then, for our Deleuzian view, capacities for action do not preexist their actualization as a self-identical possibility merely awaiting the addition of existence: there is no "grandmother cell" or single neuron—or even single preexisting network of cells—fully formed and merely awaiting activation. The Deleuzian ontological difference is not that of a possible awaiting realization but that of a virtual awaiting actualization; the virtual is composed of potentials, not possibles.[7] Second, however, the virtual helps us understand potentiality, not as self-present and simply awaiting actualization, but as a fully differentiated neuro-somatic-environmental web whose actualization proceeds by an individuation or integration of that differential field. The Deleuzian notion of a virtual potential is thus non-Aristotelian. For Aristotle, potentials are always oriented toward their telos in actuality. They are understood simply as nonactual, as that which can become actual, as in the canonical definition of capacity [hexis] at De Anima 2.1.412a22–26: "the soul is an actuality like knowledge . . . possessed but not employed." So in the famous triad, we have the ability to learn, the state of having learned, and the exercise of that which has been learned. Potentiality is oriented to form or self-presence; as Aristotle puts it elsewhere, "matter exists in a potential state, just because it may attain to its form; and when it exists actually, then it is in its form" (Metaphysics 8.8.1050a15; emphasis in the Princeton edition [Barnes 1984]). The virtual, however, Deleuze never tires of reminding us, does not resemble its actualization; there is nothing identical in its being—it is fully differential.

An example from a thinker close to the 4EA approach, Jesse Prinz, will help us see what is at stake here.[8] The ontological status of a repertoire as virtual potential can be brought to bear on Prinz's (2007, 84) treatment of the widespread notion of disposition:

> In saying that sentiments are dispositions, I don't mean to imply
> that they are not real, physically implemented states of the mind.
> As I will use the term, a psychological disposition is a standing
> state of the organism that can manifest itself as an occurrent
> state. The standing/occurrent distinction is commonly used in
> philosophy. In psychological jargon, psychological dispositions
> can usually be identified with encodings in long-term memory
> that can be retrieved by working memory and maintained there

during explicit mental processing. In neurocomputational terms, dispositions are usually identified with weighted connections between neurons that can activate the assemblies of neurons that they connect. All these ways of talking capture the basic idea that dispositions are internal states that do not always participate in information-processing, but can become active contributors under the right circumstances. A sentiment is a disposition whose occurrent manifestations (or working memory encodings, or neural activation patterns) are emotions.

With our Deleuzian lens, we can see that the standing versus occurrent scheme echoes the potential versus actual scheme, but we need not see the "standing state" or "disposition" as a self-identical state, that is, as a possible awaiting realization, or as a teleologically oriented potential awaiting actualization. Rather, we can follow Deleuze in seeing the move from disposition to occurrent manifestation as happening "under the right circumstances," that is, as an individuation of a differential field at a singular point in its progressive determination, a threshold in linked rates of change of neural firing patterns as they intersect changes in the body–world components of the entire neuro-somatic-environmental system. The singular point is set by the history of the system. The key is to think dispositions as virtual and the move to their occurrence as actualization–integration–resolution of a differential field. The "encoding" referenced by Prinz thus cannot be seen as a localized and present neural trace but as the construction of a singularity in a differential relation (linked rates of change of firing patterns) serving as the threshold for individuation and hence actualization.

A practical example complements the preceding passage and also provides a contrast with the Aristotelian notion of a self-identical potential to attain a capacity: learning to swim for Deleuze is "conjugating" the distinctive points of our bodies with singularities of the Idea of the sea to form a problematic field, a distributed and differential system of brain, body, and environment. And any one exercise of swimming is then a resolution of that problematic field, an individuation that does not resemble the virtual field but is a creative actualization of it: "this conjugation determines for us a threshold of consciousness

at which our real acts are adjusted to our perceptions of the real relations, thereby providing a solution to the problem" (Deleuze 1994,
165).[9] Note that "consciousness" here should be thought as "sentience"
rather than as full-blown reflective self-consciousness. Deleuze agrees
to a form of the mind in life thesis when he says that "larval subjects"
are coextensive with cellular life, but the adjective *larval* here shows
that he does not mean such subjects are a fully reflective self-conscious
subject (Deleuze 1994, 70–79; see chapters 8 and 9).

Let us now turn to perception, or more precisely, the relation of
acts of perception to perceptual capacity. For Deleuze, that relation
is an "intensive individuation," which can be modeled as the integration of a (virtual) differential field. To see this, note that in *Difference
and Repetition,* Deleuze interprets the Leibnizian Idea of the sea as a
system of differential relations and singularities, showing how Leibniz
helps us think conscious perception as emergent from a differential
field of tiny unconscious perceptions: the microsounds of the waves
coalescing into the murmur of the ocean (Deleuze 1994, 253; see also
Deleuze 1993, 85–100). This coalescence of microperceptions at the
threshold of consciousness is explicitly linked by Deleuze to the notion
of integration. The important thing for us is that the threshold that
determines conscious perception is not a persistent identity, a stable
property of a substantial organism, but a capacity grounded in a differential field of sensorimotor processes. To see this, we can follow
another of Deleuze's main references on perception, Bergson (1991,
46), who writes in *Matter and Memory,* "The truth is that perception
is no more in the sensory centers than in the motor centers; it mea-
sures the complexity of their relations." So we see Bergson defining
perception as the measurement of the complexity of the relations of
sensation and movement.

This is precisely the formula given by Alva Noë (2004) in his *Action
in Perception.* Noë writes, "The basis of perception, on our enactive,
sensori-motor approach, is implicit practical knowledge of the ways
movement gives rise to changes in stimulation" (8). Thus failures of
perception are due to a "breakdown in our mastery or control over the
ways sensory stimulation changes as a function of movement" (10).
Noë goes on to contrast his equation of "implicit practical knowledge"
with "mastery or control" with Kant's famous line that "intuitions

without concepts are blind" (11). As we know, Kant's theory of perceptual experience is a hylomorphic process in which formless intuitions are the material input to a production process; they are given form from transcendental sources, first by space and time as forms of outer and inner intuition, then by schematized concepts of the understanding. (The "process" here is not temporal, but follows the structure of Kant's transcendental argument.) By contrast, Noë's formulation is that what completes intuition is "knowledge of the sensorimotor significance of those intuitions." This "knowledge" is not linguaform or conceptual but is "sensorimotor bodily skill" (11). Deleuze would agree here, and the latter's notion of virtual can help us understand the ontological status of perceptual capacity as sensorimotor skill. Our perceptual capacity or sensorimotor skill is the ability to modulate the relation of the two processes of movement and sensation. As we recall, Deleuze suggests the term *virtual* for these sorts of purely differential structures. Perceptual capacity is a skill that enables us to navigate the differential elements, relations, and singularities involved in the multiplicity linking movement and sensation.

In Deleuze's terms, then, perceptual capacity is virtual, which is precisely the term Noë uses, so a comparison of the uses of this term in Deleuze and Noë is now in order. Discussing the thesis that our "impression" of the "presence and richness of the visual world is an illusion," Noë (2004, 49–50) writes, "All the detail is present, but it is only present virtually, for example, in the way that a web site's content is present on your desktop." He continues, "To experience detail virtually, you don't *need* to have all the detail in your head. All you need is quick and easy access to the relevant detail when you need it" (50; emphasis original). So here *virtual* means "accessible." Now if to be "present virtually" means that the detail is already formed, but just not in the field of vision, then this doesn't fit with Deleuze. But I do not think this is what Noë means, despite his use of the example of "a web site's content [that] is present on your desktop" (50). Rather, to be virtually present *to* an organism means the perceptual detail is not yet formed but *could be* formed by the proper manipulation of the relation between movement and sensation. The detail is potential, not possible. But it is not a preformed (Aristotelian) potential teleologically oriented to its actualization. The dative is the important clue here to our interpretation of a virtual differential potential. Perceptual detail is that which is virtually available to an organism, that which could be formed in the concrete perceptual process, versus that which is formed "in itself" and is just waiting there to enter the field of vision.

Realism and Idealism

With the notion of virtual perceptual detail, a Deleuzian and phenomenologically informed enactive cognitive science faces the classical questions of realism and idealism. In our Deleuzian-inflected 4EA approach, the worldly component of virtually available perceptual detail is not realist (the world is outside us, preformed, and we capture a picture of it and hold that picture in our heads) or idealist (the world in itself is chaotic or unknowable, so the world we experience is formed in our heads). And the visual component is neither realist (vision is a camera that will capture the preexistent information once it swings into view) nor idealist (vision creates the detail by hylomorphically informing a chaotic manifold). With Deleuze's conceptual resources added to the mix, we can say that the realism–idealism debate is based on a confusion of virtual and actual. It assumes that either the world is actual and merely awaiting capture or that our subjectivity is actual and merely awaiting raw material on which to operate with either empirical or transcendental faculties—Humean associationist regularities of human nature or the Kantian transcendental machinery. For Deleuze, the finished products, the actual world and the actual subject, are both abstractions from the concrete intensive process of experience as it integrates the virtual–differential neuro-somatic-environmental web; as we will see in a few paragraphs when we discuss the concept of affordances, the actual world and the actual subject are abstractions in the sense of being limit cases of completion for always-ongoing processes.

For both Deleuze and phenomenology, then, experience is not based in the outside or the inside. Rather, experience is in the middle, in the concrete process whose limits can be abstracted from the process and reified as actual world or actual subject. Let us first follow the phenomenological path, as it informs Noë's argument that locates experience in the middle, between subject and object. Noë (2004, 215–16; emphasis original) writes,

If the content of experience is virtual, in this way, then there is *a sense* in which the content of experience is not in the head. Nor is it in the world. Experience isn't something that happens in us. It is something we do; it is a temporally extended process of skillful probing. The world makes itself available to our reach. The experience comprises mind and world. Experience has content only thanks to the established dynamics of interaction between perceiver and world.

Our argument that a Deleuzian interpretation of the virtual status of perceptual capacity helps inform 4EA approaches gains valuable support when Noë argues against splitting experience into "an occurrent and a merely potential or dispositional aspect" (215). The potential or dispositional, as we have argued in discussing Prinz, is not preformed, self-identical, and merely awaiting realization. That Aristotelian schema is rejected by Noë, who claims that any candidate for what can become "occurrent" is itself "virtual all the way in" so that "experience is fractal and dense" (216). We might use a Deleuzian distinction at this point: the virtual detail does not "exist," but "insists," so that it "is" only as that which could be actualized out of its differential condition (Deleuze 1990, 81).

As we will shortly see, different organisms connect with different affordances even when they are based on the "same" thing. So for Deleuze, the "thing in itself" is scattered or dispersed across a virtual field of potential affordances, whose multiple actualizations depend on the relation made with an individual organism. But for all that, for all his critique of the standard picture given in the realism–idealism debate, Deleuze is not an antirealist (a major point of emphasis for DeLanda [2002]). It is just that he is a realist about Ideas, not about things in themselves (with all the appropriate caveats we gave earlier as to Deleuze's non-Platonic notion of Ideas). As he puts it in *Difference and Repetition*, "problematic Ideas are precisely the ultimate elements of nature" (Deleuze 1994, 165). So the interesting sense of realism for Deleuze is that the world has structure, but that structure is the structure of multiply realizable processes, not the structure of fully individuated things that result from those processes. This Deleuzian take can be connected to Noë's notion of enactive phenomenology:

the world has some structure, but it is not fully preformed. The world needs to be met halfway. Phenomenology is the study of the way the world reveals itself in the middle between subject and thing; it is not introspection into the picture-creating activity of an idealist subject, nor is it introspection into the camera-like abilities of a realist and representing subject. The Deleuzian notion of concrete perceptual process also operates in the middle: it is the perceiving organism that integrates the differential relations of movement and sensation and thereby individuates its perceptual objects.

The Individuation of Perceptions

We do not just have perceptual capacity as a (virtual) skill, we also have the concrete perceptual process, which for Deleuze would be an "intensive individuation" that integrates a differential field, that is, a set of linked rates of change of movement and sensation. This notion of intensity as the passage across the virtual–actual ontological difference shows us how Deleuze agrees with the phenomenological critique of realism and idealism. For Deleuze, both realism and idealism are flawed because they take one side of an opposition of finished products: a fully formed world or a fully formed subject. To avoid this, we have to move to "the genesis of real experience" (Deleuze 1994, 69) as integration of a differential field, that is, as individuation leading the actualization of the virtual. Now, for Deleuze, individuation in perception is just as much morphogenesis, that is, bringing into form, as is hurricane formation (see "Introduction I"). But perceptual morphogenesis is not hylomorphic; we are not imposing form on formless sensory matter. Rather, we are guiding the implicit forms of "sensorimotor contingencies," as Noë (2004, 105) puts it in his enactive take on Gibsonian "affordance." An affordance is in the middle, a relation of organism and environment; it is not freestanding but needs to be completed by an organism (Thompson 2007, 247).

Turning to the more complete discussion of affordances in Chemero (2009) will help us understand what is at stake here for a Deleuzian contribution to the 4EA approach. Chemero offers us a recap of what he calls "Affordances 1.1" (200) in three claims: (1) "affordances are what we perceive; they are the content of experience" (200); (2) "affordances are relations between what animals can do and features of the environment" (200); and (3) "the perception of affordances is also a relation; it is a relation between an animal and an affordance" (200).[10] Chemero's Affordances 2.0 is a dynamic theory of affordances (150–54). In developing Affordances 2.0, we are directed to start with Affordances 1.1, "then consider the interaction over time between an animal's sensorimotor abilities, and its niche, that is, the set of affordances available to it" (150). Chemero specifies two time scales here. First, we have the developmental time scale, in which an "animal's sensorimotor abilities select its niche—the animal will become selectively sensitive to information relevant to the things it is able to do" (150–51). Second, we have the behavioral time scale, in which "the animal's sensorimotor abilities manifest themselves in embodied action that causes changes in the layout of available affordances, and these affordances will change the way abilities are exercised in action. . . . Affordances and abilities causally interact in real time and are causally dependent on one another" (151).

We can give a Deleuzian reading to Chemero's notion of affordances by identifying the differential relation here. Start with the elements: affordance (Af) is the relation of animal ability (AA) to feature of environment (FE), whereas perception (P) is the relation of animal ability (AA) to affordance (Af). Thus the perception of an affordance (P of Af) is a relation (perception relates animal to affordance) to a relation (affordance relates animal ability to feature of environment). So P of Af = P of AA / FE. Dynamically speaking, however, perception is constantly changing (we are always faced with ΔP, and our sensory apparatus is that which finds the instantaneous rate of change of that change, the acceleration or deceleration). Similarly, an affordance is the link of the rate of change of what an animal can do (ΔAA) and the rate of change of features of the environment (ΔFE). So dynamically speaking, the perception of affordance (P of Af) is the relation between the rate of change of what an animal perceives (ΔP) to the linked rates of change of animal ability and features of the environment ($\Delta AA / \Delta FE$). In a formula, P of Af $= \Delta P / (\Delta AA / \Delta FE)$. In Deleuzian terms, perception is composed of differential relations and singularities; perception is an Idea. If dynamic processes are the concrete level of sensorimotor perceptually guided behavior, then the terms of Affordances 1.1 (perception, affordance, animal ability, and feature of environment) are abstractions, limit cases of freezing and reifying ongoing processes. This is why Chemero's dynamic Affordances 2.0 account is so helpful. And following this line of thought, a feature of the environment has to be something that changes at the appropriate rate for an interaction (DeLanda 2002, 90–91). For affordances to take place, there has to be a match between the time scales of animal and environment.[11] For example, sports daring and imagination can be seen as the meshing of time scales: is this rapidly closing opening between two defenders an affordance for a shot given the acceleration I might attain? You don't calculate this; you feel it as a potential you might actualize. You feel your potentials based on your history of trying similar attempts. So differentiation and integration are here practical exercises that give their results as feelings; they are not calculations that give their results as consciously accessible, numerically formed probabilities. We will see how this notion of affective and practical handling of linked rates of change pays off in our final paragraphs.

But before then, with this Deleuzian take on Affordances 2.0, we can offer two supplementary comments to Chemero's account. First, when Chemero writes that "the animal's sensorimotor abilities manifest themselves in embodied action" (151), this must be changed to read that actions "actualize the virtual ability." We know that an action has a different ontological status from an ability. But what moves us across that ontological difference between an ability and the exercise of that ability, that is, an action? It cannot just be "manifestation" because that makes abilities into preexisting individuated entities awaiting discovery, perhaps at most into possibilities awaiting the addition of existence. Rather, the move from ability to action has to be actualization as individuation in the integration of a virtual differential field. Second, the notion that embodied action can change the layout of affordances has to be read as countereffectuation, as the intensive changing the virtual conditions for future actualization. It cannot be that affordances and abilities causally interact: affordances are relations, as are abilities, and relations cannot causally interact. What can causally interact (in the sense of efficient causality) are individuated beings and acts: only the act of climbing a tree, not the unactualized ability to climb, can knock some bark off the tree or strain a muscle. It is only these individuated actions that can change the web of relations structuring the intensive processes that integrate differential fields and produce action.

Even in *E. coli* (chapter 8), we have sense-making in the living present: retention (of past differentiations) and protention (the integrated trajectory as indicating the future course of the organism). Commenting on Leibniz, Deleuze (1993) confirms our analysis of his theory of perception: "differential calculus is the psychic mechanism of perception" (90) so that "the tiniest of all animals has glimmers that cause it to recognize its food, its enemies" (92). But here we must do some interpretive work if we are to make this aspect of Deleuze relevant to 4EA cognition approaches. We must demur from the literal sense of Deleuze's remarks about differential calculus as the mechanism of perception. Differential calculus has indeed been a paradigm for computationalist models of vision (a classic example being Johansson [1976]). But for the 4EA approach, much or even most cognition is not brain-bound information processing. It is instead the direction of action of an organism in the world. The important thing, then, is the practical, embodied and embedded handling of instantaneous rates of change of changes, accelerations, and decelerations. Andy Clark cites studies of baseball players who handle a decelerating fly ball not by computation but by running so that the angle of vision to the target does not change (Clark 1999, 346, citing McBeath, Shaffer, and Kaiser 1995). The trick is to maintain the coordination of changes in the organism with changes in the environment. This practical coordination of linked rates of change has to be the sense of integration for an enactive Deleuzianism. Let us recall the notion of a "conjugation of singularities" between our bodies and the Idea of water from Deleuze's (1994, 165) analysis of swimming from *Difference and Repetition*: "our real acts are adjusted to our perceptions of the real relations, thereby providing a solution to the problem." It is this practical, embodied and embedded conjugation of singularities that provides the individuation that leads the actualization of the virtual capacity for successful engagement with the environment, the exercise of our sensorimotor perceptual capacities.[12]

PART III
Biology: Life and Mind

≈ 8 ≈

Larval Subjects, Enaction, and *E. coli* Chemotaxis

We now begin part III, in which I rehearse the relation of Deleuze and some of the currents in contemporary biology, among them the developmental systems theory of Susan Oyama and colleagues, the enactive approach of Francisco Varela and colleagues, and the eco-devo-evo approach of Mary Jane West-Eberhard (chapter 10). Chapters 8 and 9 are linked approaches to the relation of Deleuze's work and the *mind in life* position of Evan Thompson (2007).

On first reading, the beginning of chapter 2 of *Difference and Repetition,* with its talk of "contemplative souls" and "larval subjects," seems something of a bizarre biological panpsychism. Actually, it does defend a sort of biological panpsychism, but by defining the kind of psyche Deleuze is talking about, I will show here how we can remove the bizarreness from that concept. First, I will sketch Deleuze's treatment of "larval subjects," then show how Deleuze's discourse can be articulated with Evan Thompson's biologically based intervention into cognitive science, the mind in life or *enaction* position. I will then show how each in turn fits with contemporary biological work on *E. coli* chemotaxis (movement in response to changes in environment).

The key concept shared by all these discourses is that cognition is fundamentally biological, that it is founded in organic life. In fact and in essence, cognition is founded in metabolism. Thus fully conceptual recollection and recognition, the active intellectual relation to past and future—what Deleuze will call the dominant "image of thought"—is itself founded in metabolism as an organic process. This founding of cognition in metabolism can be read in an empirical sense, for just as a matter of fact, you will not find cognition without a living organism supporting itself metabolically. But it can also be read in a transcendental sense: for our thinkers, metabolism is a new transcendental aesthetic, the a priori form of organic time and space. The essential temporal structure of any metabolism is the rhythmic production of a living present synthesizing retentions and protentions, conserved conditions and expected needs. The essential spatiality of metabolism comes from the necessity of a membrane to found the relation of an organism to its environment; there is an essential foundation of an inside and outside by the membrane, just as there is an essential foundation of past and future by the living present. We thus see the necessity of a notion of biological panpsychism: every organism has a subjective position, quite literally a "here and now" created by its metabolic founding of organic time and space; on the basis of this subjective position, an evaluative sense is produced that orients the organism in relation to relevant aspects of its environment.

Let us pause for a moment to appreciate the radicality of this notion of the biological ubiquity of subjects, what we have called a *biological panpsychism*. For Deleuze in *Difference and Repetition*, the organism has an essential, albeit "larval," subjectivity based in its organic syntheses, and our active intellectual syntheses are dynamically generated from this foundation. This truly radical thesis is shared by the mind in life position. What is most interesting is that try as they might to uphold a mechanistic position in which organisms are mere "robots," the contemporary biologists we examine will also find themselves unable to avoid ascribing an essentially subjective position to single-celled organisms. Far from expecting them to experience the delight of a M. Jourdain discovering his predilection for prose, we might anticipate the shock—if not the downright dismay—of these scientists at learning that they, too, share in positing a new transcendental aesthetic, an inescapable production of a singular here and now for each organism, and the inescapable subjective production of "sense" by that organism.

In this chapter, I will concentrate on the temporal aspect of this new transcendental aesthetic and on the necessary subjectivity of the organism, as these are both treated in a manageably short text, the beginning of chapter 2 of *Difference and Repetition*. I will defer a full consideration of Deleuze's treatment of the membrane and organic spatiality until chapter 9.

Deleuze

Deleuze's overall aim in *Difference and Repetition* is to provide a "philosophy of difference," in which identities are produced by integration of a differential field (or "resolution" of a "problematic" field; the two expressions are synonymous [Deleuze 1968, 272; 1994, 211]). The philosophy of difference intersects many forms of what we might call *identitarian* philosophy, from Plato and Aristotle to Kant and Hegel and others, in which identities are metaphysically primary and differences are seen within a horizon of identity. With regard to Kantian transcendental philosophy, Deleuze attempts to replace the Kantian project of providing the universal and necessary conditions for any rational experience with an account of the "genesis" (221; 170) of "real experience [*l'expérience réelle*]," that is, the "lived reality [*réalité vécue*] of a sub-representative domain" (95; 69). As "sub-representative," such "experience" is as much corporeal and spatiotemporal as it is intellectual, as much a passive undergoing as an active undertaking. For example, the embryo experiences movements that only it can undergo (321; 249); these movements are "pure spatio-temporal dynamisms (the lived experience [*le vécu*] of the embryo)" (277; 215).

Deleuze provides two genetic accounts in *Difference and Repetition*: static and dynamic. To be fully differential, these genetic accounts must avoid a mere "tracing" of the empirical; the transcendental must be differential to never "resemble" empirical identities (176–77; 135). The more well-known of the two genetic accounts is that of chapters 4 and 5, the static genesis that "moves between the virtual and its actualization" (238; 183). Thus, instead of showing how psychological syntheses producing empirical unities are underlain by active transcendental syntheses (the categories) issued by a unified transcendental subject, Deleuze will provide a genetic account that first sets out a differential or "virtual" impersonal and preindividual transcendental field structured by Ideas or "multiplicities," that is, sets of differential elements, differential relations, and singularities (236; 182). This is the mathematical notion of differentiation, which is then coupled to the biological notion of differenciation. In this latter, complementary part of static genesis, intensive spatiotemporal dynamisms incarnate the Ideas; an intensive individuation process precedes and determines the resolution or integration of the differential Idea (318; 247). The complex notion of different/ciation, then, is the static genetic account of real experience. Again, to reinforce the connection with the mind in life school, we should recall that the passive subject undergoing experience can be an embryo: "the embryo as individual and patient subject of spatio-temporal dynamisms, the larval subject" (278; 215). Following this line of thought, by implication, Deleuze must be able to account for the genesis of the real experience of a single-celled organism; this will be our link to enaction and to current biological work.

Organic Time

Although a full treatment of Deleuze would require us to articulate the static and dynamic geneses, we will concentrate in this essay on dynamic genesis as establishing the a priori form of organic time and the necessary subjectivity of organic life. Chapter 2 of *Difference and Repetition* (Deleuze 1968; 1994) is devoted to Deleuze's work on "repetition for itself." The first step, on which we concentrate, is the discussion of the first passive synthesis of time, or habit, which produces the "living present" as the a priori form of organic time. We should note that organic time, the synthesis of habit producing the living present, is only the "foundation" of time. Deleuze's full treatment of time in *Difference and Repetition* posits a second synthesis of memory producing the pure past as the "ground" of time, while the third synthesis, producing the future as eternal return of difference, we might say, unfounds and ungrounds time.

The beginning of chapter 2 provides part of the dynamic genetic account of real experience, restricting itself, except for a brief and "ironic" remark about "rocks" (102; 75), to the biological register. It is dynamic because instead of moving from a virtual Idea to its actualization, as in static genesis, here we move from raw actuality to the virtual Idea in a series of interdependent "passive syntheses." The first section deals with only the first passive synthesis of time, the most basic or "foundational" in this dynamic genesis. To begin his genetic account, then, Deleuze must get down to the most basic synthesis; he must show how beneath active syntheses (thought) are passive syntheses (perception) and beneath passive perceptual syntheses are passive organic syntheses (metabolism). As always, the challenge is to describe passive syntheses in differential terms so as to avoid the "tracing" of empirical identities back to transcendental identities. So what Deleuze is trying to do is describe the differential transcendental structure of metabolism.

Part of the fabled difficulty of *Difference and Repetition* is Deleuze's use of free indirect discourse, in which he acts as a sort of ventriloquist for various authors (Hughes 2009). In the first section of chapter 2, Deleuze is working with Kant, Husserl, Bergson, and Hume. From Kant, we have the overall framework of transcendental philosophy (albeit in the form of a genetic account of real experience), and from Husserl, we have the notion of the lived or living present (*le présent vécu, le présent vivant* [Deleuze 1968, 97; 2004, 70]) as well as the distinction of active and passive syntheses. From Hume and Bergson, we get the notion of habit.

Syntheses are needed to join together a disjointed matter or sensation because in themselves, material or sensory instants fall outside each other: "a perfect independence on the part of each presentation . . . one instant does not appear unless the other has disappeared—hence the status of matter as *mens momentanea*" (96; 70). Deleuze goes on to distinguish three levels of synthesis of this first level of instantaneity:

1. Instantaneous presentation and disappearance: "objectively" as matter and "subjectively" as sensation
2. Passive syntheses (contraction or habit producing a living present)
   a. Organic syntheses (metabolism synthesizing matter)
   b. Perceptual synthesis (imagination synthesizing sensation)
3. Active synthesis (memory as recollection and thought as representation synthesizing perceptions)

Deleuze will distinguish the organic and perceptual syntheses by showing that organic syntheses have their own form of contraction or habit.[1] For Hume and Bergson, the psychological imagination moves from past particulars to future generalities: from a series of particulars, we come to expect another of the same kind. Deleuze will abstract the process of "drawing a difference from repetition" as the essence of contraction or habit and show that it occurs at the organic level as well as on the level of the passive perceptual imagination (101; 73).

To isolate organic syntheses as prior to perceptual syntheses (themselves prior to active intellectualist syntheses), Deleuze radicalizes Hume and Bergson. These two "leave us at the level of sensible and perceptive syntheses" (99; 72). But these syntheses refer back to "organic syntheses," which are "a primary sensibility that we *are*" (99; 73; emphasis original). Such syntheses of the elements of "water, earth, light and air" are not merely prior to the active synthesis that would recognize or represent them but are also "prior to their being sensed" (99; 73). So each organism, not only in its receptivity and perception but also in its "viscera" (that is, its metabolism), is a "sum of contractions, of retentions and expectations" (99; 73). Here we see the organic level of the living present of retention and expectation. Organic retention is the "cellular heritage" of the organic history of life, and organic expectation is the "faith" that things will repeat in the ways we are used to (99; 73). So Deleuze has isolated a "primary vital sensibility" in which we have past and future synthesized in a living present. At this level, the future appears as need as "the organic form of expectation," and the retained past appears as "cellular heredity" (100; 73).

Before we resume our treatment of the text, we can now briefly sketch the overall movement of the passage. Contraction or habit in organic syntheses is a "contemplative soul" in which we find an expectation that the next element of the same kind it has experienced will arrive. This temporal synthesis, a living present of expectation and retention, is the transcendental structure of metabolism. This move from experienced particular to expected general at the organic level is our "habit of life" (101; 74). The contemplative soul as organic synthesis or habitual contraction can also be called a "passive self" or "larval subject" (107; 78).

Now Deleuze cannot go directly to his key notion of the organic synthesis qua contemplative soul because he must first free a notion of habit from the illusions of psychology, which fetishizes activity. Psychology, by fear of introspection, misses the element of passive "contemplation." Indeed, psychology says the self cannot contemplate itself because of fear of an infinite regress of active constituting selves.[2] Deleuze's response is to pose the question of the ontological status of habit. Instead of asking how contemplation is an activity of a constituted subject, we can ask whether or not each self is a contemplation (100; 73). How do we get to habit as what a subject is rather than what it does? First, we must determine what habit does: it draws *(soutire à)* something new from repetition: difference. Habit is essentially "contraction" (101; 73). Now we must distinguish two genres of contraction: (1) contraction as activity in series as opposed to relaxation or dilation and (2) contraction as fusion of succession of elements. With the second form of contraction, we come on the notion of a "contemplative soul," which must be "attributed to the heart, the muscles, nerves and cells" (101; 74). Deleuze knows that the notion of an organic "contemplative soul" might strike his readers as a "mystical or barbarous hypothesis" (101; 74), but he pushes on: passive organic synthesis is our "habit of life," our expectation that life will continue. So we must attribute a "contemplative soul" to the heart, the muscles, the nerves, the cells, whose role is to contract habits. This is just extending to "habit" its full generality: habit in the organic syntheses that we are (101; 74).

We cannot follow all the marvelous detail of Deleuze's text, in which he discusses "claims and satisfactions" and even the question of joy, of the "beatitude of passive synthesis" (102; 74). We have to move to the question of rhythm.

In descriptions that will be echoed by the enactivists and by the contemporary biologists we will discuss, Deleuze claims that organic syntheses operate in series, and each series has a rhythm; organisms are polyrhythmic: "the duration of an organism's present, or of its various presents, will vary according to the natural contractile range of its contemplative souls" (105; 77). There are thousands of rhythmic periods that compose the organic being of humans: from the long periods of childhood, puberty, adulthood, and menopause to monthly hormonal cycles to daily cycles (circadian rhythms) to heart beats and breathing cycles, all the way down to neural firing patterns. Everything has a period of repetition, everything is a habit, and each one of these repetitions forms a living present that synthesizes the retention of the past and the anticipation of the future as need. Now "need" can be "lack" relative to active syntheses but "satiety" relative to organic passive syntheses. Deleuze writes, "Need marks the limits of the variable present. The present extends between two eruptions of need, and coincides with the duration of a contemplation" (105; 77).

Organic Subjectivity

We now have to address a change in vocabulary, as Deleuze moves toward the notion of larval subject, which will be our link to the enactivists. First, the contemplative soul becomes the "passive self," which is "not defined simply by receptivity—that is, by means of the capacity to experience sensations—but by virtue of the contractile contemplation that constitutes the organism itself before it constitutes the sensations" (107; 78). As we will see, we have to insist on the merely logical nature of this "before." But before that, one last vocabulary shift: the passive selves are "larval subjects." Of course, we cannot just replicate whole selves all the way down the organic scale. That would just be "tracing," positing identities beneath identities. Deleuze insists that "this self, therefore, is by no means simple: it is not enough to relativize or pluralize the self, all the while retaining for it a simple attenuated form" (107; 78). The larval subject is itself "dissolved," Deleuze will insist: "Selves are larval subjects; the world of passive syntheses constitutes the system of the self, under conditions yet to be determined, but it is the system of a dissolved self" (107; 78).

We might think that selves merely accompany contemplation: "There is a self wherever a furtive contemplation has been established, whenever a contracting machine capable of drawing a difference from repetition functions somewhere" (107; 78–79). But it is better to say that selves *are* contemplations. Contracting contemplations or habits or organic syntheses draw a difference from repetition. That is exactly what a self is: "The self does not undergo modifications, it is itself a modification—this term designating precisely the difference drawn [from repetition]" (107; 79). Because organic processes are serial, there is a series of such larval subjects: "Every contraction is a presumption, a claim—that is to say, it gives rise to an expectation or a right in regard to that which it contracts, and comes undone once its object escapes [*se défait dès que son objet lui échappe*]" (107; 79). This undoing of the larval subject with the rhythm of fatigue and satisfaction is the key to the notion that the self is not simple but dissolved, that is, serial and differential.

To grasp Deleuze's notion of the organism as larval subject, everything depends on how we interpret the "priority" of organic synthesis to perceptual synthesis as different levels of passive synthesis; that is, we have to interpret the term *primary vital sensibility*. What we will learn from the enactive school is that organic and perceptual syntheses are always linked in reality. The priority of organic syntheses is merely logical, for all organisms, even the most simple, have both metabolism and sensibility, or as the enactivists will put it in a phrase that will alert Deleuzians, "sense-making." We will see a reinforcement of Deleuze's merely logical "priority" of metabolism over sense-making in Ezequiel Di Paolo's distinction between autopoiesis and adaptivity. To temporarily adopt an Aristotelian vocabulary, the enactivists will show that although we can logically distinguish between them, in reality, all organisms have both a vegetative (metabolism–autopoiesis) and sensible (sense-making–adaptivity) psyche.[3]

The necessary combination of metabolic and perceptual capacities in an organic subject is a little difficult to see in *Difference and Repetition*, as Deleuze is working with the example of multicellular organisms, where metabolism and sensibility are subserved by physically distinct systems. Now even though, in multicellular organisms, we can spatially distinguish metabolic from sensory processes, we have to acknowledge internal monitoring, a "sensing" of the state of organism—or better, a synthesis (i.e., a differentiation–integration) that establishes the trajectory of the system: where a process is going and with what acceleration. In any event, Deleuze wants to expose thousands of contemplative souls or "little selves" as thousands of organic syntheses "before" passive perceptual syntheses and active intellectual syntheses (which Kant unifies in a subject via the transcendental unity of apperception). Deleuze's strategy is thus reminiscent of Nietzsche seeing a multiplicity of drives beneath the illusory unified ego.

But does Deleuze's emphasis on multiplicity mean he treats the organism as an "illusion"? It all depends on how we interpret the following phrase from the preface to *Difference and Repetition*. Discussing the "generalized anti-Hegelianism" that is "in the air nowadays [*dans l'air du temps*]" (1; xix; translation modified), Deleuze writes, "The modern world is one of simulacra. . . . All identities are only simulated, produced as an optical 'effect' by the more profound game [*jeu*] of difference and repetition" (1; xix). Is this Deleuze writing in his own name, setting out his thesis, or is it a report of what is in the air? Is an organism only an "illusion"? Whatever we might finally say about the unity of the organism in *Difference and Repetition*—although I briefly return to the issue in the conclusion, I will defer that full reading for now—we can at least say that our task is made more difficult by the lack of an explicit discourse on the membrane, which does not appear until the following year's *Logic of Sense*. Nonetheless, by the time we reach the straightforwardly realist and materialist stance of *A Thousand Plateaus*, it is clear that organismic stratification is not an illusion. Strata are real ("a very important, inevitable phenomenon that is beneficial in many respects" [Deleuze and Guattari 1987, 40]) and valuable ("staying stratified is not the worst thing that can happen" [161]). Conversely, with a long enough time scale, we can see that although organisms are not illusions, they are only temporary patterns, diachronically emergent patterns unifying multiple material processes for a time. This does not prevent us from articulating Deleuze and enaction; the emphasis on synchronic emergence—on the necessary systematic functioning of metabolism—in autopoiesis as the essential structure of living things could never deny the death of individuals (see "Introduction II").

What is radical about Deleuze's strategy is that by following its logic, this underlying multiplicity is true for unicellular organisms as well. Deleuze pluralizes even unicellulars, both synchronically (metabolism and perception are separate processes) but also diachronically. Every iteration of a process, each case in a series of organic syntheses, is a contemplative soul—each has its own rhythm, and it is the consistency of those rhythms that allows the cell to live. Death, we can speculate, occurs when the rhythms of the processes no longer mesh. Shifting musical terms, we can say that life is harmonious music; death is disharmony. On the supraorganismic scale, death as disharmony is the condition for creativity, for the production of new forms of life, new processes.[4] But on the organismic scale, though we can also affirm disharmony as the condition of creativity, a prudent experimentation is called for: "Dismantling the organism never meant killing yourself" (Deleuze and Guattari 1987, 160).

So even though we must be literal when we say the "living present"—it occurs on the organic level "before" it occurs on the perceptual and intellectual levels—we have to remember that this priority is merely logical; in all real organisms, organic synthesis is always accompanied by perceptual syntheses. In each organism, multicellular or unicellular, the synchronic emergent unity of the organism is always an achievement, a unification of many "little selves." But there is diachrony here as well; for Deleuze, each little self is never fully present to itself but is "dissolved" in a series of repetitions of its process. The key is to describe this dissolved or multiple or differential biological psyche without falling into a needless projection of unified active or intellectualist synthesis onto it; that is, the key is to describe passive synthesis as a logically distinct but really linked series of multiple organic and perceptual syntheses. In doing so, we will have isolated the level of the organic "larval subject" and will have thereby defined the multiple levels of Deleuze's "biological panpsychism."

To summarize, then, the passive self is never fully self-present because the passive organic and perceptual syntheses on which active syntheses are built are differential in three aspects. Each passive synthesis is serial (there is never one synthesis by itself but always a series of contractions, each with its own rhythmic period); each series is related to other series in the same body (at the most basic level, the series of organic contractions is linked to that of perceptual contractions as these are related to those of motion: echoing the enaction school, we can say that all perception is sensorimotor); and each series is related to other series in other bodies, which are themselves similarly differential (the series of syntheses of bodies can resonate or clash). Together the passive syntheses at all these levels form a differential transcendental field within which subject formation takes place as an integration or resolution of that field; in other words, even at this most basic level, larval subjects are the patterns of these multiple and serial syntheses, which fold in on themselves (again, a full treatment of the issue would demand that we articulate the role of the membrane), producing a site of lived and living experience, spatiotemporal dynamism and sentience or minimal awareness, a "primary vital sensibility."

Enaction

Although the emphasis on difference for Deleuze and on autonomy for the enactivists makes them somewhat strange bedfellows, the notions of "primary vital sensibility" and of the "larval subject" we have just traced in *Difference and Repetition* can let us see some significant resonances between the two discourses with regard to organic time and organic subjectivity. For the first aspect, organic time, we will concentrate on Jonas (2003), and for the second, on Di Paolo (2005); both of these are woven into the argument of Thompson (2007).

Organic Time

The enactivists straightforwardly talk of the new transcendental aesthetic we found in Deleuze as "biological time and space" (Thompson 2007, 155, citing Jonas 2003, 86). We find this expressed as a living present found in the simplest of organisms, a synthesis of retention and protention (Jonas 2003, 85–86). Furthermore, need is as rhythmic and affective for the enactivists as it is for Deleuze. Thompson (2007, 156) writes, "Concern, want, need, appetition, desire—these are essentially affective and protentional or forward-looking."

Let us turn to Jonas's (2003, 64–92) magnificent essay "Is God a Mathematician? The Meaning of Metabolism" for more detail on these notions; we will see the same first steps of a dynamic genesis (from instantaneity to the living present) here as in *Difference and Repetition*. Jonas proposes to test, against the case of the organism, the modern claim that God is a mathematician (65). First, Jonas reviews the history of that notion, from Plato's *Timaeus* through Leibniz. What distinguishes the ancient and modern treatments of nature is the algebraic treatment of motion on the part of the moderns (67). Thus with the moderns we find "analysis of becoming" rather than "contemplation of being"; for the moderns, it is process as such, rather than its perfection in an end state, that is the object of knowledge (67). This mathematical change of method, when applied to physics, means that "the functional generation of a mathematical curve becomes the mechanical generation of the path of a body" (68).

Here is the key for us, the connection with Deleuze's reaching the starting point of dynamic genesis in the *mens momentanea* (Deleuze 1968, 96; 1994, 70). For Jonas, modern mathematical physics gives us time as a series of instants such that the physical states of a process are externalized, one to the other: "each of them determined anew by the component factors operative at that very instant" (Jonas 2003, 68). Such fragmentation means that analysis meets no resistance; in other terms, there is no wholeness, only an aggregation of moments, and so ontological emergence is denied: "rationality of order . . . must be explained by reference to the . . . most elementary types of event . . . their singleness alone is the basically real, and the 'wholeness' of their conjoint result is an appearance with no genuine ontological status" (69).

We cannot treat all the riches of the historical sections of Jonas's text, as he moves from a reading of the *Timaeus,* where the Demiurge is needed to redeem the passivity of matter (70), to modern materialism and its dualistic counterpart, idealism, a shift that results in an inversion to which we have become inured: "'Matter' in fact, in the sense of 'body,' becomes more rational than 'spirit'" (73). This entails that "not only the mindless but also the lifeless has become the intelligible as such," a standard that means the moderns must understand life starting from "dead matter" (74).

Passing now to his interrogation of the purely mathematical physical analysis of metabolism (in other words, testing the reduction of biology to physics), Jonas proposes the wave as the physicist's model of complex physical form, a form that is wholly reducible to an aggregate. The wave, as an "integrated event-structure," has no ontologically emergent status; "no special reality is accorded that is not contained in, and deducible from, the conjoint reality of the participating, more elementary events" (77). Furthermore, Jonas adds, what is true of the wave must be true of the organism as object of divine intellection. Without need of the "fusing summation of sense," for God, "the life process will then present itself as a series, or a web of many series, of consecutive events concerning these single, persisting units of general substance" (77). Once again, we find physical time as a pure self-exteriority, as a series of instants.

For Jonas, however, such a reductive account misses the ontological emergence that makes of life an "ontological surprise," and the organism a system, a "unity of a manifold." The organism is "whole" as "self-integrating in active performance," an "active self-integration of life" (79). The "functional identity" of organisms relative to the materials it metabolizes is constituted "in a dialectical relation of *needful freedom* to matter" (80; emphasis original). Both elements, need and freedom, constitute the "transcendence" of life, and this transcendence constitutes a living present, a metabolically founded transcendental aesthetic or a priori form of organic time: "self-concern, actuated by want, throws open . . . a horizon of time . . . the imminence of that future into which organic continuity is each moment about to extend by the satisfaction of that moment's want" (85). For Jonas, in a way that highlights the partiality of Deleuze's treatment in *Difference and Repetition*, organic space is founded by organic time: an organism "faces outward only because, by the necessity of its freedom, it faces forward: so that spatial presence is lighted up as it were by temporal imminence and both merge into past fulfillment (or its negative, disappointment)" (85).

Jonas then draws the consequences for the question of the adequacy of purely mathematical physics for the phenomenon of life; in other words, he shows the necessity of a dynamic genesis from instantaneity to the living present: "with respect to the organic sphere, the external linear time-pattern of antecendent and sequent, involving the causal dominance of the past, is inadequate" (86). With life on the scene, "the extensive order of past and future is intensively reversed" so that the determination of "mere externality" by the past has to be supplemented by the recognition that "life is essentially also what is going to be and just becoming" (86).

Organic Subjectivity

Even with the notion of the "primary vital sensibility" of the larval subject of organic syntheses as our guiding thread, pairing Deleuze and enaction still seems odd. Developing out of the autopoiesis school founded by Humberto Maturana and Francisco Varela, the enactive position worked out by Evan Thompson (2007) in *Mind in Life* seems too focused on autonomy and identity to be usefully paired with Deleuze's philosophy of difference. Although autopoietic theory, developed in the 1970s at the height of the molecular revolution in biology, performed an admirable service in reasserting the need to think at the level of the organism, it is clear that autopoiesis is locked into a framework that posits an identity horizon (organizational conservation) for (structural) change. For autopoietic theory, living systems conserve their organization, which means their functioning always restores homeostasis; evolution is merely structural change against this identity horizon. Now even if Deleuze ultimately does not think the organism is an "illusion," when it comes to "life," he stresses the creativity of evolution over against the conserved identity of the organism; thus, for Deleuze, the organism is "that which life sets against itself in order to limit itself" (Deleuze and Guattari 1987, 503). Nonetheless, strictly with regard to the "primary vital sensibility" of the organism we have seen in *Difference and Repetition,* Deleuze and enaction can be brought together when we follow how Thompson supplements the undoubted emphasis on identity preservation of autopoiesis with a more dynamic and differential concept of "adaptivity" drawn from the work of Ezequiel Di Paolo. With this addition, we can see the possibility of a more fruitful interchange with Deleuze.

The key is to recognize that autopoiesis entailed not just organizational maintenance but cognition or sense-making. For Maturana and Varela, autonomous systems have sufficient internal complexity and feedback that "coupling" with their environment "triggers" internally directed action. This means that only those external environmental differences capable of being sensed and made sense of by an autonomous system can be said to exist for that system, can be said to make up the world of that system (Maturana and Varela 1980, 119). The positing of a causal relation between external and internal events is only possible from the perspective of an "observer," a system that itself must be capable of sensing and making sense of such events in *its* environment (81). So with autopoiesis, the autonomous system is always linked to its environment, and organization provides an identity horizon for structural change. But autopoiesis is only sufficient for maintenance of identity. To account for sense-making, Thompson turns to Ezequiel Di Paolo: "A distinct capacity for 'adaptivity' needs to be added to the minimal autopoietic organization so that the system can actively regulate itself with respect to its conditions of viability and thereby modify its milieu according to the internal norms of its activity" (Thompson 2007, 148).

With this important connection in mind, we can move to consider sense-making. Witness the single-celled organism's ability to make sense. "Sense" has, perhaps fittingly, a threefold sense: sensibility, signification, and direction.[5] A single-celled organism can sense food gradients (it possesses sensibility as openness to the environment), can make sense of this difference in terms of its own needs (it can establish the signification "good" or "bad"), and can turn itself in the right sense for addressing its needs (it orients itself in the right direction of movement). This fundamental biological property of affective cognition is one reason why the Cartesian distinction of mental and material has no purchase in discussions of sense-making. There is no "mental" property (in the sense of full-blown reflective consciousness) attributable to the single-celled organism, but because there is spontaneous and autonomous sense-making, there is no purely "material" realm in these organisms either. The enactive claim is that affective cognition in humans is simply a development of this basic biological capacity of sense-making.

Turning now to Di Paolo's essay, we see that he distinguishes within Maturana and Varela's work the all-or-nothing character of organizational maintenance from a more dynamic notion of homeostatic regulation. "Whereas homeostasis connotes the existence of active mechanisms capable of managing and controlling the network of processes that construct the organism, conservation is a set-theoretic condition that may or may not be realized in an active manner. It merely distinguishes between changes of state without loss of organization and disintegrative changes" (Di Paolo 2005, 435).

For Di Paolo, organizational conservation cannot explain organismic sense-making—directed action responding to environmental change relevant for the organism—precisely because it is all-or-nothing: "But what makes bacteria *swim up* the gradient? What makes *them* distinguish and prefer higher sugar concentrations? As defined, structural coupling is a conservative, not an improving process; it admits no possible gradation" (437). Di Paolo insists that an organism's sense-making, its judgment as to the improvement of conditions relative to its need, is beyond the scope of autopoiesis: "Even if the current rate of nutrient intake is lower than the rate of consumption (leading to certain loss of autopoiesis in the near future), bacteria will not seek higher concentrations *just because* they are autopoietic since improving the conditions of self-production is not part of the definition of autopoiesis" (437).

The key for us is to see that adaptivity requires a dynamic emergent self unifying a multiplicity of serial processes. We might say that autopoiesis entails synchronic emergence, whereas adaptivity entails diachronic emergence. Notice the dynamic monitoring of multiple processes Di Paolo isolates here as necessary for generating singular norms of each organism: "Only if they are able to monitor and regulate their internal processes so that they can generate the necessary responses anticipating internal tendencies will they also be able to appreciate graded differences between otherwise equally viable states. Bacteria possessing this capability will be able to generate a normativity *within* their current set of viability conditions and *for themselves*. They will be capable of appreciating not just sugar as nutritive, but the direction where the concentration grows as useful, and swimming in that direction as the right thing to do in some circumstances" (437).

Adaptive mechanisms (the measurement of the trajectory of the system against a norm and the regulative means of bringing deviations back to that norm—or indeed of changing the norm itself) are serial, and so the emergent self of the organism is in Deleuze's terms a "system of a dissolved self" (Deleuze 1968, 107; 1994, 78). In general, we have to stress the "systematic" nature here to see the connection of Deleuze with adaptivity, but the dissolution of serial selves is clear when Di Paolo (2005, 444–45; emphasis original) writes that "the operation of single adaptive mechanisms is in normal circumstances self-extinguishing but their interaction, the ongoing coupling with the environment, and the precariousness of metabolism, make their collective action also self-renewing, thus naturally resulting in *valenced rhythms of tension and satisfaction*."

So we might want to relate the "simple self" of Deleuze to the all-or-nothing character of autopoiesis and the "system of a dissolved self" to the dynamic character of adaptivity. That is, in adaptivity, there is a measuring of the trajectory of the organism against norms ("anticipating internal tendencies"). For it to be the continual monitoring and regulation of an ongoing organism in its life span, that measurement has to be serial, that is, rhythmic, dynamic, and constantly renewable ("self-extinguishing"). It cannot just be abstract "structural change" as opposed to "organizational maintenance." Deleuze is going to call each snapshot of a dynamic series of modifications, each "drawing of a difference from repetition," the "larval subject." The seriality of such a subject is indicated by the fact that the self "comes undone [*se défait*] once its object escapes" (Deleuze 1968, 107; 1994, 79); this is the "self-extinguishing" of a "single adaptive mechanism" for Di Paolo.

E. coli Chemotaxis

We have brought Deleuze and enaction together, at least from a certain perspective. But what if neither discourse relates to contemporary biology? To ground the discussion, we will look at the description of organic and perceptual syntheses in *E. coli* chemotaxis, a favorite example of sense-making for the enactivist school, in two recent biology works, Howard Berg's (2004) *E. coli in Motion* and Dennis Bray's (2009) *Wetware*. We will look at two aspects of their work to make the connection with Deleuze and with enaction: first, their account of synthesis as differentiation–integration, as "drawing a difference from repetition," that is, their establishment of a transcendental aesthetic for organic life, the living present as retention and protention, a constantly renewed "here and now"; second, their fear of organic subjectivity coupled with their inability to forgo first-person evaluative language.

Organic Time

We will find here the Deleuzian notion of passive synthesis as constituting the living present. Our authors stress the temporality of perception for their objects of study. Bray stresses the retentive aspect of *E. coli,* who "continually reassess their situation" by means of "a sort of *short-term memory*" (Bray 2004, 7; emphasis original). Such "bacterial memory" can be tested by exposing them to a step change in the concentration of an attractant: "Now it is clear that what the bugs respond to is not the concentration of aspartate per se but its rate of change" (94). Bray interprets these results in terms that cannot fail to delight any reader of Deleuze. "But once aspartate has settled down to a steady concentration, the bug no longer responds. Biologists call this adaptation, but a mathematician examining the time course of response would call it differentiation. By measuring the rate of change in the signal, the receptor cluster has in effect performed calculus!" (94). In other words, the bacterium has repeated its measurement of aspartate and drawn a difference from that repetition: it has performed a differentiation.

But the living present is a synthesis of retention and protention. Berg's work on temporal synthesis reveals the protention aspect as well as the insightful character of Deleuze's treatment of contractile habit as "drawing a difference from a repetition." Berg (2004, 57) first clearly shows retention as one aspect of the passive synthesis of the living present: "to correct its course, the cell must deal with the recent past, not the distant past." But then we see that the living present is serial, that it draws a difference from a repetition; Berg writes that "to determine whether the concentration is going up or down, the cell has to make two such measurements and take the difference" (57). Berg shows that this perceptual synthesis is temporal rather than spatial; describing the results of a key experiment, he writes, "The response to the positive temporal gradient was large enough to account for the results obtained in spatial gradients" (36). So the cell repeats its sampling procedure (it analyzes the environment, breaking it down to identify the concentration of molecules of interest) and then synthesizes the two results. What we see here in this passive synthesis is differentiation (calculation of the instantaneous rate of change of a gradient) and integration (calculating the trajectory of the change by combining the results of previous differentiations). We thus have sense-making in the living present: retention (of past differentiations) and protention (the integrated trajectory as indicating the future course of the organism).

In further confirmation of the Deleuzian and enactivist treatments of the living present, these passive syntheses are rhythmic. Owing to its being buffeted by the Brownian motion of water molecules, after about ten seconds, an *E. coli* cell "drifts off course by more than 90 degrees, and thus forgets where it is going" (49). The living present has limits to its retention; it has an essential "forgetting." Continuing with his analysis, Berg writes, "This sets an upper limit on the time available for a cell to decide whether life is getting better or worse. If it cannot decide within about 10 seconds, it is too late" (49–50). Just as it has an upper limit to its living present, "a lower limit is set by the time required for the cell to count enough molecules of attractant or repellent to determine their concentrations with adequate precision" (50). More precisely, "diffusion of attractants or repellents sets a lower limit on the distance (and thus the time) that a cell must swim to outrun diffusion (to reach greener pastures), as well as on the precision with which the cell, in a given time, can determine concentrations" (56). As Berg puts it, "if it is to go far enough to find out whether life is getting better or worse, it must outrun diffusion" (56). This minimal time for perceptual synthesis is one second, "approximately equal to the mean run length" (56). With Berg's analyses of *E. coli* chemotaxis, we see here a constantly renewed living present, the constitution of a singular here and now for each bacterium.

Organic Subjectivity

In his preface, Bray (2009, ix) writes that he received a rejection note from another publisher accusing him of writing a book about "single-celled organisms possessing consciousness." Bray reacts indignantly, but we will see that he protests too much in writing that "single cells are not sentient or aware in the same way that we are. To me, consciousness implies intelligent awareness of self and the ability to experience introspectively accessible mental states. No single-celled organism or individual cell from a plant or animal has these properties" (ix). No one, least of all Deleuze and the enactivists, would complain of this perfectly defensible high bar to meet for the ascription of "consciousness." But Bray has thrown "sentience" and "awareness" in too quickly with "consciousness," as we can see when he calls *E. coli* "robots." Bray writes that "an individual cell, in my view, is a system that possesses the basic ingredients of life but lacks sentience. It is a robot made of biological materials" (ix). The "robot" as line of defense against accusations of biological panpsychism is repeated by Howard Berg, who also writes, regarding his "top down, or outside in" treatment of cell populations, that from this perspective, *E. coli* should be seen as "robots programmed to respond to external stimuli" (Berg 2004, 19).

To avoid the charge of a too easy ascription of microsubjectivity, Bray (2009) takes a strong computationalist and representionalist stance. "It is as though each organism builds an image of the world—a description expressed . . . in the language of chemistry" (x). The most intense locus of this representation is found in the genome and protein synthesis: "From a time-compressed view, the sequences and structures of RNA, DNA, and proteins can be thought of as continually morphing in response to the fluctuating world around them" (x). Thus we come to the "central thesis of the book—that living cells perform computations" (xi). So, to avoid any hint of biological panpsychism, for Bray, cells are nonsentient robots.

Once we enter the book, however, we find Bray bothered about mechanisms missing something. "Like manic pathologists at an autopsy competition, we have littered our workbenches with the dissected viscera of cells. . . . But where in this museum of parts do we find sensation, volition, or awareness? Which insensate substances come together, and in what sequence, to produce sentient behavior?" (5). However, Bray soon returns to his computationalist position: "The molecular mechanism of E. coli chemotaxis is a superb illustration of cellular information processing" (6). But he cannot sustain the mechanistic position. Owing to Brownian motion from buffeting by water molecules, "to pursue any direction for more than a second or so, bacteria have to constantly reassess their situation" (7). But if it is *their* situation, they must have a proper point of view—it is not just "the" situation but "their" situation. We can all see here the instability of this discourse, its shifting from third-person to first-person perspective.

In his discussion of the mechanism of that reassessment, Bray is worried about subjectivity. "Words like *memory, awareness,* and *information* are easy to use but require careful definition to avoid misunderstanding. I'm using *short-term memory* here in a colloquial, nonspecialist way, referring to how a swimming bacterium carries with it an impression of selected features of its surroundings encountered in the past few seconds" (7; emphasis original). But despite these qualifications, he has to return to the first-person perspective. Adding aspartate to a solution will take the percentage of tumbling cells from 20 percent to near zero. This is because "the cells have experienced an improvement in their environment (a taste of food) and consequently persist in their current direction of swimming" (7). "Experienced" here shows the inevitability of some notion of minimal subjectivity.

We see the same instability of discursive stances in Howard Berg. He first seems to indicate the necessity of a first-person perspective in his distinction between "aesthetics" and "material gain." He writes that the modern era of E. coli research begins in the 1960s, when "Julius Adler demonstrated that E. coli has a sense of taste, that is, that bacterial chemotaxis is a matter of aesthetics rather than material gain" (Berg 2004, 15). In discussing such sampling, though, Berg reverts to a third-person perspective: "Adler was able to show that *E. coli* responds to chemicals that it can neither transport (take up from surrounding medium) nor metabolize (utilize as a source of energy or raw material)" (24). Another example is perhaps more telling. Berg writes of "attractants" and "repellants," which seemingly imply a first-person perspective, but he defines them in purely third-person behavioral terms: "chemicals whose gradients strongly affect the motile behavior of wild-type *E. coli*" (25, Table 3.1).

Much as they try, however, in the long run, the authors cannot avoid a blend of third-person mechanism and first-person evaluation. Bray (2009, 7) writes of how *E. coli* possesses "a sort of short-term memory that tells them whether conditions are better at this instant of time than a few seconds ago. By 'better' I mean richer in food molecules, more suitable in acidity and salt concentration, closer to an optimum temperature." The seemingly innocuous term "food" is the giveaway, for *food* is a relational term: sucrose is only food "for" an organism; it is not food in itself (Thompson 2007, 158). And clearly "suitable" and "optimum" are relative to the life process of organisms.

A final example from Bray, linking retention in the living present to subjective evaluation, follows: "Bacteria store a running record of the attractants they encounter. This tells them whether things are better or worse: whether the quantity of food molecules in their vicinity is higher or lower than it was a few seconds ago" (Bray 2009, 94). Here, again, we see the mixture of mechanistic (third-person) and evaluative (first-person) language. If a "quantity" of (chemical) "molecules" is being measured, we have a third-person description of a mechanism, but if it is "food" being measured, we have a first-person perspective; the measurement of food is relative to the need of an organism. The inevitability of first-person evaluative terms is clear soon when Bray writes, "It's a pragmatic strategy: if conditions are improving, continue swimming; if not, tumble and try another direction" (94).

Let us conclude this all-too-brief discussion of the treatment of organic subjectivity in contemporary biology by returning to Berg, who is somewhat more straightforward in his adoption of evaluative terms and a first-person perspective. In discussing the run versus tumble behavior of individual cells, Berg writes that "*E. coli* extends runs that are favorable (that carry cells up the gradient of an attractant) but fails to shorten runs that are not (that carry cells down such a gradient . . .). Thus, if life gets better, *E. coli* swims farther on the current leg of its track and enjoys it more. If life gets worse, it just

relaxes back to its normal mode of operation. *E. coli* is an optimist"
(Berg 2004, 35).

Deleuze and Enaction

We cannot exaggerate the fit of enaction and Deleuze. We have
stressed the serial, dynamic, affective, and differential character of
enaction, but we have underplayed some of Deleuze's radicality.

To have a full picture of the notion of organism in *Difference and
Repetition*, we would have to discuss it in terms of static genesis, for the
organism is one of the prime examples of Ideas, first discussed in terms
of Geoffrey Saint-Hilaire and anatomical elements and then updated
in terms of genetics (Deleuze 1968, 239–40; 1994, 184–85). But Ideas
are incarnated by spatiotemporal dynamisms, which are processes of
individuation, so a confrontation with Deleuze's reading of Simondon
will be necessary (317; 246). The larval subject is the individual in the
process of individuation and hence is tied to a metastable field in an
ongoing process of "transduction." The priority of individuation over
differenciation must be respected (318; 247), and this leads Deleuze
to a prescient critique of genetic determinism: "The nucleus and the
genes designate only the differentiated matter—in other words, the
differential relations which constitute the pre-individual field to be
actualized; but their actualization is determined only by the cytoplasm,
with its gradients and its fields of individuation" (323; 251; see chapter
10 for more detail on this point).[6]

Even on the basis of this brief sketch, it might appear, then, that
the emphasis in enaction on the notion of an autonomous system
overemphasizes the individual as a self-conserving product as opposed
to individuation as an always ongoing process. From this perspective,
the embryo as paradigmatic "larval subject" is merely a more intense
site of individuation than the adult; however sclerotic and habitual,
the adult is only the limit of the process of individuation; it is never
actually reached; no more than the virtual does the actual exist rather
than insist. In terms of autopoietic synchronic emergence, then, we
might say that enaction relegates the metastable field to a coupled
environment and limits transduction to metabolism, while in terms
of adaptivity's diachronic emergence, it neglects ontogenesis in favor
of adult function and restricts transduction to homeostatic regulation.

I am under no illusions as to my capacity at the present time to prove these assertions; I merely wish to record them as speculations to be pursued in future work.

Finally, we should note that by radicalizing what we might call the Bergsonian and Whiteheadean threads, which intersect the Simondonian thread, we can see a total panpsychism in *Difference and Repetition* that surpasses the biological (see chapter 10 for more on this notion). Deleuze notes that the mathematical and biological notions of differentiation and differenciation employed in the book are only a "technical model" (Deleuze 1968, 285; 1994, 220). Now if "the entire world is an egg" (279; 216), then every individuation is "embryonic," we might say, even "rocks" (282; 219) and "islands" (283; 219). Now if rocks and islands as individuation processes are embryonic, then they, too, have a psyche: "every spatio-temporal dynamism is accompanied by the emergence of an elementary consciousness" (284; 220). We will not pursue this line of thought but will note that by the time of *Anti-Oedipus* (Welchman 2009) and *A Thousand Plateaus* (Bonta and Protevi 2004), Deleuze and Guattari explicitly thematize that the syntheses are no longer bound to "experience," however widespread, but are fully material syntheses, syntheses of nature in geological as well as biological, social, and psychological registers. With this full naturalization of syntheses, the question of panpsychism is brought into full relief because now syntheses are syntheses of things and of experience.

≈ 9 ≈

Mind in Life, Mind in Process

This chapter examines the idea of "biological space and time" found in Evan Thompson's (2007) *Mind in Life* and in Deleuze's (1994) *Difference and Repetition*. Tracking down this "new Transcendental Aesthetic" intersects new work done on panpsychism. Both Deleuze and Thompson can be fairly said to be biological panpsychists. That is what "mind in life" means: mind and life are coextensive; life is a sufficient condition for mind.[1] Deleuze is not just a biological panpsychist, however, so we will have to confront full-fledged panpsychism. At the end of the chapter, we'll be able to pose the question whether we can supplement Thompson's "mind in life" position with a "mind in process" position and, if so, what that supplement means both for his work and for panpsychism.

Philosophical linkages often fall into the "strange bedfellows" category, and though for some, the linking of cognitive science, biology, and phenomenology in Thompson's *Mind in Life* is quite strange enough, this chapter seeks to show a further unexpected connection between Thompson's work and a poststructuralist classic, Deleuze's *Difference and Repetition*. In particular, we will examine the idea of "biological space and time" found in Thompson (2007, 154–57; see also Jonas 2003, 86) and Deleuze (1994, 70–79).

To begin, let us notice that the mind in life position continues the twentieth-century trend of displacing human, language-expressed, top-level, reflective rational consciousness as the sole or prime or most basic or most important candidate for cognition, a position that would have corporeal practical engagement as a privative form, as sloppy or distorted or approximate theory. We see this displacement in the phenomenological tradition, in Husserl's analyses of passive synthesis and the life world, as well as in Heidegger and Merleau-Ponty, where embodied practical engagement is primary. In this displacement of rational reflective conscious thought, we also see the connection with

Deleuze, who picks up the post-Kantian demand, explicit in Salomon Maimon, for genesis rather than mere conditioning (see the new translation of Maimon's [2010] *Essay on Transcendental Philosophy*; for commentary, see Jones [2009] and Smith [2009]). Rather than the Kantian deduction of conditions of human rational reflective consciousness, Deleuze holds that we have to show the genesis of real experience from within experience. The post-Kantian challenge relayed by Deleuze is to show how space and time, categories, and Ideas develop in "dynamic genesis," starting with the sheer atomic exteriority of sensations to one another (what Deleuze [1994, 70] will call *mens momentanea*) and moving to "virtual Ideas" (see "Introduction I").

The key concept shared by Deleuze and Thompson is that the sort of cognition for which Kant posited his transcendental conditions develops from a fundamentally biological cognition, what Thompson calls *sense-making*. The mind in life claim is that fully conceptual recollection and recognition, the active intellectual relation to past and future, is founded in sense-making or *enactive cognition* (what Deleuze will call the linkage of perceptual and organic syntheses). Mind in life, read as coextensivity, means that life is fundamentally autopoiesis and that cognition is fundamentally sense-making. Thus *mind in life* means "autopoietic sense-making," or the control of the action of an organism in its environment. Sense-making here is threefold: (1) sensibility as openness to environment, (2) signification as positive or negative valence of environmental features relative to the subjective norms of the organism, and (3) direction or orientation the organism adopts in response to 1 and 2.

This founding of human cognition in enactive cognition entails a new transcendental aesthetic, the a priori but always concrete genesis of organic time and space. The essential temporal structure of any organism is the rhythmic production of a living present synthesizing retentions and protentions, conserved conditions and expected needs. The essential spatiality of any organism comes from the necessity of a membrane to found the relation of an organism to its environment; there is an essential foundation of an inside and outside by the membrane, just as there is an essential foundation of past and future by the living present. We thus see the necessity of (at least) a notion of biological panpsychism: every organism has a subjective position, quite literally a here and now created by its founding of organic time and space; on the basis of this subjective position, an evaluative sense is produced that orients the organism in relation to relevant aspects of its environment.

Enactive Space and Time

In the decentering of reflective consciousness we sketch here, we see three moves: two familiar, the other being the novelty of the mind in life positions. First is the familiar phenomenological move of showing how high-level thought, exemplified in science, is a transformation of the life world or of Dasein's practical involvement (depending on whether you prefer a Husserlian or Heideggerian vocabulary). This is a synchronic shift of position within adulthood: adults are not first and foremost scientists in everyday life; they are instead practically and corporeally engaged with the world. In other words, we have to show how know-that (science) is a transformation of corporeal space-time (a reformed Transcendental Aesthetic) and corporeal know-how (a reformed Transcendental Analytic). As Donn Welton (2000, 299) puts it, for Husserl, "the objects that we do find in Kant's Analytic, full-blown objects of science, belong to a higher order and are not experientially basic. Constitution at this higher level must be understood not as elementary but as a transformation of what is elemental."

Second is another familiar move: genetic phenomenology. At least, it is familiar now thanks to the efforts of Donn Welton, Anthony Steinbock, and other "New Husserlians," who have mined the archives containing Husserl's manuscripts (Steinbock 1995; Welton 2000; 2003). Here we have to trace the development of corporeal space-time and corporeal know-how from embryo to adult, that is, along the developmental or ontogenetic time scale. This is where we get a first reformulation of the Transcendental Aesthetic. In *The Other Husserl*, Donn Welton shows how the Transcendental Aesthetic is renamed in Husserl: instead of the Kantian opposition of sensibility and understanding (judgment), we have Husserl's opposition of experience and judgment (understanding). Because we have passive synthesis in what Kant would have as merely passive sensibility, there is a noematic sense in perception, prior to active understanding–judgment, and these passive syntheses include associative, kinesthetic, and time-consciousness syntheses (Welton 2000, 298). Directly addressing Husserl's genetic undercutting of Kant's Transcendental Aesthetic, Welton writes,

> At yet a deeper and final level of genetic analysis Husserl discovers that space and time themselves are not just "forms" but are generated, on the one hand, by the interplay of position, motility, and place, and on the other, by the standing-streaming flow of the process of self-temporalization itself. Husserl's studies of the self-generation of space and time are clearly the most difficult of all his genetic studies. (254)

Our key question, then, is whether dynamic genesis or Husserlian genetic phenomenology is restricted to the ontogenetic time scale, that is, the development from embryo to adult.

If it is so restricted, then we need a new name for the third move, which is the key novelty of the mind in life position: we have to do "evolutionary" genetic phenomenology (and not just ontogenetic). In Deleuze's terms, we have to do dynamic genesis on the evolutionary time scale. That is to say, we have to show how single-celled organisms generate their own concrete space and time (a biological Transcendental Aesthetic) as well as display sense-making (a biological Transcendental Analytic) *and* how this develops along the evolutionary time scale into the potentials for what will develop along the human developmental time scale, that is, genetic phenomenology proper as the constitution of corporeal space-time and corporeal know-how, from embryo to adult. And then, finally, we can trace the synchronic transformation of corporeal space-time and categories–Ideas into science or human "high reason."

Now if we can have a genetic phenomenology on the evolutionary time scale—if "evolutionary genetic phenomenology" makes sense—then we have to talk about its basis, an empathy condition based on our living experience. To address the sense-making or affective–cognitive "metabolic situation" of the single-celled organism, we ourselves have to be living beings. First, here is Jonas (2003):

> On the strength of the immediate testimony of our own bodies *we* are able to say what no disembodied onlooker would have a

cause for saying: that the mathematical God in his homogeneous analytical view misses the decisive point—the point of life itself: its being self-centered individuality, with an essential boundary dividing "inside" and "outside." (79; emphasis original)

Thompson (2007) agrees:

> Empathy is a precondition of our comprehension of the vital order, in particular of the organism as a sense-making being inhabiting an environment. . . . [A] bodily empathy . . . widened beyond the human sphere to ground our comprehension of the organism and our recognition of the purposiveness of life [Thompson here refers to Husserl, *Ideas II* and *Ideas III*]. Empathy in this sense encompasses the coupling of our human lived bodies with the bodies of other beings we recognize as living, whether these be human, animal, or even—particularly for biologists with a "feeling for the organism" [Thompson here refers to Evelyn Fox Keller's biography of Barbara McClintock]—bacteria. (165)

Using this empathy condition to explore the experience of even the simplest living beings, Thompson and Jonas straightforwardly talk of a new transcendental aesthetic as "biological time and space" (Thompson 2007, 155, citing Jonas 2003, 86). We find this expressed as a living present found in the simplest of organisms, a synthesis of retention and protention in the concrete form of metabolism and need (Jonas 2003, 85–86).

For Jonas, a physicomathematical account misses the ontological emergence that makes of life an "ontological surprise" and of the organism a system, a "unity of a manifold." (We will return to the question of emergence and panpsychism.) The organism is "whole" as "self-integrating in active performance"; it is an "active self-integration of life" (Jonas 2003, 79). The "functional identity" of organisms relative to the materials they metabolize is constituted "in a dialectical relation of *needful freedom* to matter" (80; emphasis original). Both elements, need and freedom, constitute the "transcendence" of life, and this transcendence constitutes a living present, a metabolically founded transcendental aesthetic or a priori form of organic time: "self-concern, actuated by want, throws open . . . a horizon of time . . . the imminence of that future into which organic continuity is each moment about to extend by the satisfaction of that moment's want" (85). For Jonas, echoing Heidegger's grounding of Dasein's spatiality in its temporality in *Being and Time* #70, organic space is founded by organic time: an organism "faces outward only because, by the necessity of its freedom, it faces forward: so that spatial presence is lighted up as it were by temporal imminence and both merge into past fulfillment (or its negative, disappointment)" (85). Jonas then draws the consequences for the question of the adequacy of purely mathematical physics for the phenomenon of life; in other words, he shows the necessity of a dynamic genesis from instantaneity to the living present: "with respect to the organic sphere, the external linear time-pattern of antecedent and sequent, involving the causal dominance of the past, is inadequate." With life on the scene, "the extensive order of past and future is intensively reversed" so that the determination of "mere externality" by the past has to be supplemented by the recognition that "life is essentially also what is going to be and just becoming" (86).

Deleuze, Simondon, and Membranes

We discussed Deleuze's treatment of organic time in chapter 8. We now move to consider Deleuze's reliance on Gilbert Simondon's analysis of organic space-time as found in the relation of metabolism and membrane. As we have seen with Jonas, the essential spatiality of metabolism comes from the necessity of a membrane to found the relation of an organism to its environment; there is an essential foundation of an inside and outside by the membrane, just as there is an essential foundation of past and future by the living present. The interest of the new biological Transcendental Aesthetic is to see its intertwining of space and time in the relation of membrane and metabolism.

Prior to the publication of *Mille Plateaux* in 1980, Deleuze only mentions biological space founded by membranes a few times, always with reference to Simondon. So let us turn to the section of Simondon's *L'Individu et sa genèse physico-biologique* titled "Topologie et ontogénèse" (Simondon 1995, 222–27) to discover what he says about how membranes and metabolism entail a biological Transcendental Aesthetic.[2] The basic concept of Simondon's work is the process of individuation or "transduction" starting from a metastable field.[3] In Simondon's work, a metastable field does not contain individuals; it is preindividual but poised for individuation. Simondon's usual figure for transduction or individuation from a metastable field is crystallization: in the supersaturated field, there are gradients of density but no crystalline forms nor crystals as individuated entities. There is the potential for crystallization, made actual when provoked by an external shock. From a metastable field, a process of individuation allows for the distinction of an ever-processual individual and milieu. Individuation as "transduction" is thus an always ongoing maintenance of metastability between individual and milieu.

Let us follow Simondon's treatment of biological space-time, the new biological Transcendental Aesthetic, in *L'Individu*. To establish the singularity of the living being *(le vivant)*, "it would be necessary to exhibit [*produire*] the topology of the living being, its particular type of space, the relation between a milieu of interiority and a milieu of exteriority" (Simondon 1995, 223). The key is that the new biological Transcendental Aesthetic is topological, not Euclidean. We cannot be fooled by the seemingly Euclidean or "absolute" inside–outside in single-celled organisms,[4] for "the essence of the living being is perhaps a certain topological arrangement that cannot be known on the basis of physics and chemistry, which utilize in general a Euclidean space" (223). While it is the case that there is an "absolute" inside–outside of the single-celled organism, it is not a Euclidean spatiality but the dynamic and topological maintenance of metastability that counts:

> For this organism, the characteristic polarity of life is at the level of the membrane; it is in this region [*à cet endroit*] that life exists in an essential manner as an aspect of a dynamic topology which itself maintains the metastability by which it exists. (224)

So we see how it is the dynamic topological process of individuation that constitutes biological space-time. The interior is the accumulated past, the exterior the forthcoming future. Concerning the relation of interiority and the past, Simondon writes, "The entire mass of living matter which is in the interior space is actively present to the exterior world at the limit of the living being: all the products of the past of the individuation [*de l'individuation passée*] are present without distance and without delay" (225). While interiority constitutes the past, exteriority constitutes the future: "The fact that a substance is in the milieu of exteriority means that that substance can come forth [*peut advenir*], be proposed for assimilation, [or] wound [*léser*] the living individual: the substance is to come, is futural [*est à venir*]" (225).

The full contours of the new biological Transcendental Aesthetic come into focus as past and future combine in a living present constituted by the membrane:

> At the level of the polarized membrane, the interior past and the exterior future face one another [*s'affrontent*]: this face-off [*affrontement*] in the operation of selective assimilation is the present of the living being [*le présent du vivant*], which is made up of this polarity of passage and refusal, between substances which pass into the past [*substances passées*] and substances which come forth futurally [*adviennent*], [substances which are] present [*présentes* used here as an adjective] one to the other by means of [*à travers*] the operation of individuation. (Simondon 1995, 226)

However, we must never reify the membrane: it is the process of individuation maintaining a dynamic topology that constitutes the new Transcendental Aesthetic of living present as relation of interior and exterior, past and future: "the present is that metastability of the relation between interior and exterior, past and future; it is in relation to this allagmatic activity of mutual presence that the exterior is exterior and the interior is interior" (226).

To conclude this brief treatment, we can note that Simondon is quite clear that the new, biological Transcendental Aesthetic he articulates in his philosophy of transductive individuation is a departure from Kant: "Topology and chronology are not a priori forms of sensibility, but the very dimensionality of the living being as it individuates itself [*la dimensionnalité du vivant s'individuant*]" (Simondon 1995, 226).

Life and Creativity

The following section is something of a departure from a strict focus on space and time, but as the topic is so important, let us consider Simondon's (1995) definition of life, which is quite close to that of autopoiesis:

Life is self-maintenance [*auto-entretien*] of a metastability, but a metastability that requires a topological condition: structure and function are linked, for the most primitive and profound vital structure is topological. (224)

Simondon's definition is quite close to the definition of life in autopoiesis, but there are some notable differences. The similarity comes from the notion of self-maintenance of a topological dynamics in which structure and function are linked. But the "metastability" thematized by Simondon is an interesting twist. The binary logic of autopoiesis—conservation or dissolution—had to be supplemented by the dynamic notion of "adaptivity" developed by Ezequiel Di Paolo (2005) and explicated by Thompson in *Mind in Life*. The reason for this supplement is that autopoiesis is only sufficient for the maintenance of the organism's self-identity. To account for sense-making, Thompson turns to Di Paolo's notion of adaptivity: "A distinct capacity for 'adaptivity' needs to be added to the minimal autopoietic organization so that the system can actively regulate itself with respect to its conditions of viability and thereby modify its milieu according to the internal norms of its activity" (Thompson 2007, 148).

But what about Simondon's "metastability"? Can this term, discussed in terms of virtuality by Deleuze, be covered by "adaptivity"? It would take more time than we are able to devote to it here, but we can pose a few points for further development. The key for us is to see that adaptivity requires a dynamic, emergent self unifying a multiplicity of serial processes. We might say that autopoiesis entails synchronic emergence, whereas adaptivity entails diachronic emergence. Notice the dynamic monitoring of multiple processes Di Paolo (2005, 437; emphasis original) isolates here as necessary for generating singular norms of each organism:

Only if they are able to monitor and regulate their internal processes so that they can generate the necessary responses anticipating internal tendencies will they also be able to appreciate graded differences between otherwise equally viable states. Bacteria possessing this capability will be able to generate a normativity *within* their current set of viability conditions and *for themselves*. They will be capable of appreciating not just sugar as nutritive, but the

direction where the concentration grows as useful, and swimming
in that direction as the right thing to do in some circumstances.

The comparison of enactive biology and Deleuze is complicated, however, by Deleuze's notion of intensive individuation processes. Deleuze
is a process philosopher, one focused on creativity and novelty (Shaviro
2009). We can truly say that autopoiesis is not a substance concept,
at least insofar as substance is seen as reified, for what is conserved
in autopoiesis is not the organism as stable thing but the organism
as self-maintaining membrane–metabolism recursive process. But
what of the notion of creativity in life on which Deleuze focuses?
Does the autopoietic organism help us think about life's creativity?
For Deleuze and Guattari, the answer is no; the organism is actually
only the resting point between bursts of eco-devo-evo creativity: "the
organism is that which life turns against itself to limit itself" (Deleuze
and Guattari 1987, 503). So "life" for them is most fully displayed in
evolutionary creativity, even if, in their more sober moments, Deleuze
and Guattari admit, perhaps even grudgingly, that the organism or
autopoietic conservation is the condition for another round of biological creativity: "Dismantling the organism has never meant killing
yourself. . . . Staying stratified . . . is not the worst that can happen"
(160–61). After all, dead men tell no tales, and dead organisms produce
no creative variations.

What are we to make of all this talk of creativity? Is this just another barbarous or mystical hypothesis? Far from being a vitalist fantasy,
as we will see in chapter 10, Deleuze's emphasis on ontogenetic and
evolutionary creativity echoes the notion of "developmental plasticity" developed by Mary Jane West-Eberhard (2003; see also Pigliucci
2010). Although I cannot show it in detail here, I would claim Deleuze
is a multi–time scale thinker, an eco-evo-devo thinker: along with
"involution" (what Lynn Margulis calls "symbiogenesis"),[5] he would
agree with West-Eberhard that creativity in developmental plasticity
provides a source of the variation with which evolution by natural
selection works, other than the canonical source, random mutation.[6]
But autopoiesis and adaptivity seem limited to the behavioral time
scale. Even granted that autopoiesis is a (self-focused) process term,
we might say that the notion of the autonomous system overemphasizes stability, whereas a Deleuzian–Simondonian transductive

individuation, even if it doesn't emphasize creativity per se, at least provides the conditions for it. From this perspective, the embryo as paradigmatic "larval subject" is merely a more intense site of individuation than the adult; however sclerotic and habitual, the adult is only the limit of the process of individuation. There is always the chance for change, for development of new patterns. Of course, they have to fit within limits of viability, as autopoiesis insists, but it is a matter of emphasis: autopoiesis with its emphasis on conservation and adaptivity with its emphasis on homeostasis versus Deleuze's emphasis on creativity, for which Simondon's notion of transductive maintenance of metastability serves as its condition. In terms of autopoietic synchronic emergence, then, we might say that enaction relegates the metastable field to coupled environment and limits transduction to metabolism, while in terms of adaptivity's diachronic emergence, it neglects ontogenesis in favor of adult function and restricts transduction to homeostatic regulation.

Spatiotemporal Dynamisms

To this point, we have discussed the new biological Transcendental Aesthetic. But Simondon's notion of individuation extends below the organic level; transductive individuation is prebiotic as well as biotic. There are important dynamic topological differences between crystallization and organic individuation, but "there might be an intermediary order of phenomena, between parcellary microphysics and the macrophysical unity of the organism; this order would be that of genetic processes, chronological and topological, that is to say, the processes of individuation, common to all orders of reality in which an ontogenesis operates" (Simondon 1995, 227).

Let's spend a minute on the fascinating difference between crystals and organisms as Simondon articulates it:

> Vital individuation does not come *after* physico-chemical individuation, but during this individuation, before its completion, by suspending it at the moment when it has not attained its stable equilibrium. . . . The living individual would be in some manner, at its most primitive levels, a crystal in the state of being born [*à l'état naissant*], amplifying itself without stabilizing itself. (150; emphasis original)

Simondon appeals to neoteny (slowing down) to explain this idea. So within the organic realm, we also see individuation as the suspension of metastable processes. In a startling image, the animal is the "inchoate plant,"

> developing and organizing itself by conserving the motile, receptive, and reactive possibilities which appear in the reproduction of vegetative life [*la reproduction des végétaux*]. . . . Animal individuation feeds on [*s'alimente*] the most primitive phase of vegetative individuation, retaining in itself something anterior to the development of the adult plant [*végétal adulte*], and in particular maintaining, during a longer time, the capacity of receiving information. (150)

These prebiotic "genetic processes," operating by means of a deferral of stability or maintenance of metastability, are what Deleuze calls "spatio-temporal dynamisms." In his terms, they are intensive processes rather than virtual structures or actual products.

Let's turn to Deleuze's (2004) essay "The Method of Dramatization," which has a somewhat more clear presentation than *Difference and Repetition*. Spatiotemporal dynamisms "create particular spaces and times," in a non- or prebiotic Transcendental Aesthetic:

> Beneath organization and specification [the actual], we discover nothing more than spatio-temporal dynamisms: that is to say, agitations of space, holes of time, pure syntheses of space, direction, and rhythms. (96)

Spatiotemporal dynamisms, as intensive processes of impersonal individuation with their own space-time, entail a second new Transcendental Aesthetic, this time non- or prebiotic. Although individuation is a general case, covering the prebiotic, Deleuze finds biology a better model than Simondon's crystallization. But biology is only a model for Deleuze's notion of intensive processes that actualize a virtual Idea (Deleuze 1994, 220–21). So, when he unleashes one of his most infamous gnomic utterances, "the whole world is an egg" (Deleuze 2004, 96; see also Deleuze 1994, 251), we cannot restrict the extension of spatiotemporal dynamisms to the biological realm. In other words, it is not the "egg" we should concentrate on but rather "the whole world." For transductive individuation in all registers, organic, physical, and social, one needs a preindividual field with virtual potentials that are not individuated, that do not "resemble" the products produced by intensive individuation processes structured by those potentials (Toscano 2006).

Making the connection to the new Transcendental Aesthetic pursued by our thinkers, we see that Deleuze (2004, 99) will claim that "what I am calling a drama [another term for spatiotemporal dynamism] particularly resembles the Kantian schema" (see also Deleuze 1994, 216–17). Seeing spatiotemporal dynamisms as the analogue to schematisms is linked to the post-Kantian demand for genesis of the Transcendental Aesthetic, that is, space and time as generated rather than posited as conditions: "We would have to distinguish what belongs to space and what to time in these dynamisms, and in each case, the particular space-time combination. Whenever an Idea is actualized, there is a space and a time of actualization" (Deleuze 2004, 111). To locate the space and time of actualization, we must first distinguish three registers: virtual, intensive, and actual (DeLanda 2002; Bonta and Protevi 2004). The intensive is the space-time of individuation processes, that is, actualization of the virtual. In the virtual register, we have virtual space: the meshed continuum of Ideas with zones of indiscernibility between them. And we have virtual time as progressive determination of Ideas, the "movement" from determinable but undetermined differential elements, their reciprocal determination in differential relations—as these are themselves determined by individuation processes[7]—and the ideal of complete determination in the singularities these relations generate.[8] Deleuze calls virtual space "depth," or *spatium,* and virtual time "Aion."

In the intensive register, we find intensive spatial processes: foldings, cascades, convection currents, and so on. With intensive time, we see the time of thresholds and critical points, the time of kairos. Finally, extensive or actual space has universal measurements, millimeters, meters, and so on, whereas extensive–actual time is similarly universally measurable with the same units: the time of Chronos or clock–calendar time. The difference in intensive space-time and extensive space-time is the existence of "critical" points and moments in the former: the moment of a process reaching a threshold that produces a qualitatively new pattern is not just any old moment or "time T1," just as the point at which currents bend is not just any old spatial point at specific values of coordinates x, y. Rather, critical times and spaces are immanently determined *as* critical in the intensive process that unfolds with its own concrete space and time; it cannot be compared in a universal framework to some other moment or point.

Deleuze's spatiotemporal dynamisms are found in multiple registers: (1) in the physical register, the spatial density gradients and temporal critical points are reciprocally determining in crystallization; (2) in the ontogenetic organic register, cellular displacement and temporality of gene expression networks are linked in embryonic development; and (3) in the evolutionary organic register, the distribution of plastic developmental systems (multiplicity of concrete space and time of ontogenesis in a population) provides the variation for the temporality of genetic accommodation in Mary Jane West-Eberhard's work. The contrast, then, of concrete intensive space-time dynamisms and abstract universal extensive time is the contrast between the Transcendental Aesthetic of Deleuze and Simondon and that of Kant, in which universal space and time are the a priori forms of intuition.

The Question of Panpsychism

When we realize that each spatiotemporal dynamism for Deleuze has a larval subject, we are forced to tackle the question of panpsychism. Although he uses many biological examples in *Difference and Repetition*, they are only examples of spatiotemporal dynamisms and larval subjects. As we will see, rocks and islands are spatiotemporal dynamisms, too, so they too will have a "larval subject"! Deleuze (2004, 97) writes,

> Dynamisms are not absolutely subjectless, though the subjects they sustain are still only rough drafts, not yet qualified or composed, rather patients than agents, only able to endure the pressure of an internal resonance or the amplitude of an inevitable movement. A composed, qualified adult would perish in such an environment. The truth of embryology is that there are movements which the embryo alone can endure: in this instance, the only subject is larval.

One of the great advantages of the mind in life position is that it enables us to escape from the badly posed Cartesian problem of the relation of the mental and the physical. But then we have a problem with the emergence of life and mind: the move from the abiotic to the biotic, from the noncognitive to the cognitive. And with this move, we come on the question of panpsychism.

Recent work has gone back to the problem of panpsychism (Skrbina 2005; 2009). We will very briefly touch on two of the contemporary approaches to panpsychism Skrbina identifies, information theory of the "cybernetic mind" type and process philosophy (Skrbina 2005, 246). We'll begin with process philosophy. To Cartesian mechanists, panpsychism is laughable, if not maddening—the abject of thought. Panpsychism is inconceivable: extended substance is dead, inert, unconscious, nonsentient, bereft of experience. But Thompson (2007, 440–41) rejects the Cartesian extended substance picture of nature in favor of a radical processualism:

> In the context of contemporary science . . . "nature" does not consist of basic particulars, but fields and processes. . . . There is no bottom level of basic particulars with intrinsic properties that upwardly determines everything else. Everything is process all the way "down" and all the way "up," and processes are irreducibly relational—they exist only in patterns, networks, organizations, configurations, or webs. . . . There is no base level of elementary entities to serve as the ultimate "emergence base" on which to ground everything.

Insofar as the major process philosopher of the twentieth century, Whitehead, was a panpsychist, we have at least a prima facie invitation to pursue the connection of Thompson's radical processualism and panpsychism (for a recent piece on Whitehead and panpsychism, see Basile [2009]). With the coextension of mind and life, we come to the question by the panpsychists, is mind in life too restrictive with its definition of mind? If there is no mind prior to life, then mind must radically emerge with life. But with the process view, there is no *radical* emergence from a baseline of elementary particles. There is emergence in the sense of new structures generating new capacities, but the panpsychist would say that these new capacities are the development of potentials in the "lower level." When cognitive capacities are at stake, a panpsychist would say that mind gets more complex as we find life, but it doesn't radically emerge with life. A Cartesian radical emergentist would say that there is dead unmindful matter that when properly arranged becomes living and minded. But is that really less strange than the panpsychist position? In fact, Strawson (2006) will say that radical emergence is no better than "magic," so it is actually a more rigorous position to be a panpsychist.

Let us concentrate on Thompson. In *Mind in Life*, Thompson examines several cases at the borderline of autopoietic cognition, starting with Stuart Kauffman's autocatalytic loops (Thompson 2007, 104–5). Recall the mind in life position that strongly links autopoiesis and cognition: "cognition is behavior or conduct in relation to meaning and norms that the system itself enacts or brings forth on the basis of its autonomy" (Thompson 2007, 126). On this reading, you need the physical instantiation of metabolism–membrane–metabolism recursivity to have an autonomous subjectivity such that life and organismic sense-making are linked. Thus Thompson will rule out Kauffman's autocatalytic loops as the basis or minimal example of life: they don't have a recursive membrane–metabolism structure, so they don't have autonomy and don't enact a subjective position (Thompson 2007, 105). Later he examines other borderline cases: the tessellation automaton of Bourgine and Stewart and the autocatalytic miscelles and viscelles of Bitbol and Luisi (125). While Thompson leaves undecided whether these systems are minimal cases of autopoiesis or only proto-autopoietic systems, with the strong autopoietic definition of cognition given earlier, they, too, fail to qualify as autonomous autopoietic systems because they do not produce a physically realized membrane–metabolism recursivity and hence an autonomous subject position (125–27).

But what about a simpler definition of cognition? There is information transfer and self-organization in autocatalytic loops, and this fits the cybernetic definition of mind offered by Gregory Bateson (1972, 460) when he identifies "mind as synonymous with cybernetic system—the relevant total information-processing, trial-and-error completing unit."[9] Deleuze has such a cybernetic model in his notion of the "dark precursor" (see "Introduction I").

Beyond that, if we push it, we can see a total panpsychism in *Difference and Repetition* that surpasses the biological to the level of spatiotemporal dynamisms or the self-organizing cybernetic mind level. Deleuze (1994, 220) notes that the mathematical and biological notions of differentiation and differenciation employed in *Difference and Repetition* are only a "technical model." Now if "the entire world is an egg" (216), then every individuation is "embryonic," we might say, even rocks: "On the scale of millions of years which constitutes the time of their actualization, the hardest rocks in turn are fluid matters which flow under the weak constraints exercised on their singularities" (219). Now if rocks and islands as individuation processes are embryonic, then they, too, have a psyche: "every spatio-temporal dynamism is accompanied by the emergence of an elementary consciousness" (220). By the time of *Anti-Oedipus* and *A Thousand Plateaus*, Deleuze and Guattari explicitly thematize that the syntheses they investigate are fully material syntheses, syntheses of nature in geological as well as biological, social, and psychological registers (Welchman 2009). Not just organic syntheses, but inorganic ones as well, are spatiotemporal dynamisms. With this full naturalization of syntheses, the question of panpsychism is brought into full relief because material syntheses are as much syntheses of experience as they are syntheses of things, as we see in the title of chapter 3 of *A Thousand Plateaus*: "The Geology of Morals: Who Does the Earth Think It Is?"

We thus have a second new Transcendental Aesthetic here with Deleuzian spatiotemporal dynamisms. It is the Transcendental Aesthetic of larval subjects, of mind in physical self-organizing processes, echoing Bateson's cybernetic mind. So the question is, how do we relate this to *Mind in Life*? Can we have a coherent defensible notion of mind that's broader than that of sense-making of an autopoietic organism, one based on information transfer and self-organization in physical processes (crystallization, convection currents, lightning, hurricanes)? Then the question of emergence of mind is pushed down below emergence of life. How far down? Is there a point of emergence we can locate? That's what the panpsychists deny. For them, it is mind all the way down. Thompson will say it is process all the way down, but he does not say whether there is a nonautopoietic notion of mind that accompanies process. Is there a "mind in process" position we need to think about? Thompson's subtitle is "Biology, Phenomenology, and the Sciences of Mind." Is there a "Physics, Phenomenology, and the Sciences of Mind" book to be written?

To move toward a conclusion, let us note that a classic objection to panpsychism is based on a worry about the overuse of the principle of parsimony: we cannot push parsimony too far because the fewer principles we have, the more we risk stretching them beyond their useful extension. So we have to worry that a definition of mind as mere information transfer involved in self-organization is so broad as to be meaningless: if convection currents in a pot of boiling water are mind, what good is such a broad definition? But conversely, what's exciting about dynamic systems modeling is that it shows self-organizing processes in an extremely wide range of registers, from convection currents through neurodynamics. So if self-organization is a univocal concept, that is, if there is a nontrivial shared structure between convection currents and neurodynamics, then we have identified a fundamental principle that links the inorganic and organic registers. So we are back to the cybernetic challenge: is information transfer and self-organization capable of being called "mind" in a defensible fashion? It would not be autopoietic cognition because it doesn't involve a membrane–metabolism recursive process and hence an autonomous subject position. But would not it be "mind in process," even if it is not "mind in life"?

To conclude somewhat abruptly, if there is mind in process, that is, mind all the way down, just as there is process all the way down, that means we really have our work cut out for us in discussing this second new Transcendental Aesthetic, the non- or prebiotic one. It is not that we don't have enough to talk about with a biological Transcendental Aesthetic, but if we want to follow Deleuze all the way, we'll have to go not only "beyond the turn" in (human) experience, as Bergson (1991, 185) puts it, but "beyond the turn" of (living) experience, out into the "plane of consistency" we find posited in A Thousand Plateaus.

⇒ 10 ⇐

The Virtual Status of "Unexpressed Genetic Variation"

This concluding chapter is the newest and most ambitious effort of the book.[1] I will argue that Deleuze's ontological framework can illuminate two key concepts—"unexpressed genetic variation" and "genetic accommodation"—in Mary Jane West-Eberhard's (2003) *Developmental Plasticity and Evolution*.[2] I will first show how the strong antigenetic reductionist stance of some strands of contemporary biology reveals an interlocking system of genetic and epigenetic factors guiding development. In this new perspective, the genome is no longer a blueprint determining development with epigenetic factors being merely occasions for regulatory genes to kick into action and orchestrate development. Rather, the genome is something akin to a musical score from which musicians draw and recombine bits and pieces while leaving a track of their actions, thus in effect rewriting the score in the very playing of it (Keller 2000, 63).[3] This distributed developmental guidance system, with both genetic and epigenetic factors being active participants, can be seen as a differentiated virtual multiplicity in Deleuze's terms. Finally, I will show that West-Eberhard's concept of genetic accommodation is an example of Deleuze's countereffectuation. In other words, genetic accommodation is the way intensive individuation processes bring forth previously unexpressed virtual potentials; it is the way individuating development changes the differentiated multiplicity guiding that very development's next steps.

The term *developmental plasticity* in West-Eberhard's title refers to the way in which organisms of the same species can develop differently in response to different environments. *Genetic accommodation* entails that some environmentally induced adaptive phenotypic change is then stabilized by natural selection. Finally, *unexpressed genetic variation* is that on which developmental plasticity and genetic accommodation rest. Unexpressed genetic variation is inherited but was previously hidden from selection by developmental robustness, the counterweight to plasticity; given similar environments, organisms in the same species but with different genomes are nevertheless similar in form and behavior. In sum, genetic accommodation is the way in which environmental induction of novel phenotypic traits—resulting from developmental plasticity, in which new gene expression networks call on previously hidden, that is, unexpressed, genetic variation—becomes genetically encoded (West-Eberhard 2003, 145, 499ff.).

What West-Eberhard suggests is that even if new gene expression networks are constructed from "unexpressed genetic variation," that is, they exist as constructed from the potentials inherent in that unexpressed variation, they can nonetheless "sometimes precede the evolution of the lineal sets of contiguous exons that characterize many [hereditary] genes" (329). In other words, rather than function following heredity, it precedes it and leads to its consolidation. As we will see, the conceptual point that brings Deleuze and West-Eberhard together is that the potentials of the unexpressed genetic variations to undergird new gene expression networks are virtual. This means that they are revealed only retrospectively, after irreducibly singular experiments in developmentally plastic individuation; these experiments select the differential relations of genetic and epigenetic factors in what Deleuze calls the "progressive determination" of the virtual in the process of actualization.

Development and Heredity

The mainstream view of contemporary biology combines two historically distinct, but now merged, streams of thought, evolution and development.[4] The evolutionary thought stream includes—and in some cases is seen as exclusively coextensive with—the notion of evolution by natural selection, which has three factors: variation (from random mutation), heredity (by DNA), and selection (organisms adapting to changing environments). The developmental thought stream merged with the evolutionary in the notion of gene-determined development, in which a contiguous string of nucleic acids codes for a string of amino acids in proteins that then determine phenotypic traits. By now, however, this once-settled viewpoint is the subject of vigorous debate, with Piglucci and Müller (2010) presenting the case for the utility of an "extended synthesis" that considerably nuances these ideas: the sources of variation are now thought to include developmental plasticity (and for Lynn Margulis, "symbiogenesis"); heredity and development are now seen to have epigenetic components; and selection needs to consider "niche-construction" and "co-evolution."

In this section, I am going to follow Evelyn Fox Keller's (2000) presentation in *The Century of the Gene,* where she shows that it is the very progress of molecular biology that has undermined the gene-determined development picture.[5] The standard picture was popularized in the slogan "DNA makes RNA. RNA makes protein. Proteins make us." But the real process is more complicated than this unilinear process:

1. DNA in nucleus is separated (its two strands pull apart).
2. An enzyme (RNA polymerase) *transcribes* the bottom strand in complementary mRNA (messenger RNA).
3. The primary mRNA transcript is transported out of the nucleus into the cytoplasm.
    a. The introns (intervening or "junk" DNA stretches) are excised.
    b. The exons (expressing or protein-coding DNA stretches) are spliced together.
4. On the ribosome, the tRNA (transfer RNA) binds to mRNA by recognizing triplet codons on the mRNA.
5. The tRNA adds an amino acid monomer, correlative to the triplet codons of the mRNA, to the protein polymer chain under construction. This is the process known as *translation*.
6. The protein chain, when complete, drops off the ribosome and goes on to play its role in the cell.

Note that between steps 3 and 4, complex processes of splicing and editing go on. Thus the *primary* mRNA transcript at step 3 has to be edited and spliced to form the *mature* mRNA transcript that goes to the ribosome in step 4. Why these extra steps? It turns out that proteins sometimes need separated strands of DNA for their synthesis. Often there are chunks of inactive or "junk" DNA between the strings of active DNA. Let me pause here and define some technical terms. Inactive

DNA is known as *introns* for "intragenic region" or "intervening sequence," whereas active DNA is known as *exons* or *expressed DNA*. Exons can also be called *structural* genes as opposed to *regulatory* genes; the latter control the expression of structural genes instead of coding for proteins, as structural genes do. It is actually even more complicated than this, as some stretches of DNA are expressed, but with seemingly no functional effect (this is known as *ectopic expression*; see note 2).

So to return to our narrative of the complexities of gene expression, the introns have to be cut out (edited) from the primary mRNA transcript, and the exons have to be strung together (spliced) to form the mature mRNA transcript.[6] Another twist: exons can be spliced together in different orders in a process called *alternative splicing*. This means that more than one mature mRNA transcript can be produced from the same primary mRNA transcript (i.e., from the same DNA string). Thus there is no longer a one-to-one correspondence of DNA sequence and synthesized protein. We have now one gene (DNA string) = many (mRNA transcripts) = many proteins. We can think of it this way: we have to learn to separate the *hereditary gene* (the contiguous string of DNA passed down in reproduction) from the *functional gene* (the end product of the transcription process forming, from separated strings of DNA, a gene that plays a role in protein synthesis).

The crucial question is the following: what controls the editing and splicing that regulates the construction of functional genes from hereditary genes? *It depends on the state of the cell at any one time.* Thus control has migrated from DNA (structural plus regulatory genes) to the complex system in which DNA plays a (certainly very important) role, but no longer a controlling role. But the story is not over yet. Not only are different proteins formed from the "same" hereditary gene, but proteins function in different ways, according to the cellular context in which they find themselves. This change in protein function due to changes in their structure is known as *allostery*. So now we have, instead of one protein = one function, the case that one protein = many functions.

In this way, in following Fox Keller's presentation of the changes in the received view, we have gone from one string of DNA = one gene = one protein = one function to one string of DNA (structural/hereditary gene) = many (functional) genes (many mature mRNA transcripts) = many proteins = many functions. Of course, the first equation is an ideal case: everyone always acknowledged the possibility of errors at each stage (i.e., errors in transmission of DNA in heredity, then transcription or translation errors). Still, the point is that the classic reference point was always a linear, self-contained process by which phenotypic traits could be understood as reducible to proteins produced by genes as DNA strings. The upshot is that we have to give up the idea that "genes code for traits" and be satisfied with the notion that individual mature mRNA transcripts code for individual proteins (Wheeler 2007). But we cannot go simply from hereditary genes (DNA strings) to functional genes (individual mature mRNA transcripts), nor can we go from proteins to traits.

In this all-too-brief tour, we have seen how gene expression depends on cell dynamics that are part of larger networks. (In fact, even the stability of the hereditary gene or DNA sequence is influenced by external events. In what is called *stress-induced mutagenesis,* mutation rates increase in crisis situations [Jablonka and Lamb 2005, 320–23]; this capacity has itself evolved in what is called the evolution of *evolvability* [353].) But let us talk about development before we talk any more about heredity; we have to understand both to understand West-Eberhard's views on developmental plasticity as leading the way for evolutionary change.

So far we have spoken about individual cell metabolism and linear DNA sequences. But the process of development includes cell differenciation: we have many different types of mature cells.[7] Thus gene expression has to follow a temporal pattern, even though hereditary genes are linear strings of DNA. At first, Keller writes, hopes were high that Jacob and Monod's operon model, which depended on a distinction between regulatory and structural genes, could account for gene expression. In this picture, development could be controlled from inside the genome by a "genetic program" whose operations were merely occasioned by cell conditions (Keller 2000, 93–101). But now biologists acknowledge the active rather than merely occasioning role that epigenetic factors play in heredity and development. What are these epigenetic factors? Eva Jablonka and Marion Lamb's (2005) *Evolution in Four Dimensions* presents the case for a far-reaching characterization. (Note that although Jablonka and Lamb present their case in terms of heredity, all these factors are also involved in development [West-Eberhard 2007, 441].) Working our way outward from DNA, we note that it is packaged and coiled on the chromosomes. This packaging, DNA and chromosomal proteins together, is called *chromatin*, which plays an important role in gene expression. Next we find the cytoplasm, which, in earliest development, is the fertilized egg. The gradients in the egg turn out to be very important in development. There are also many connections between cytoplasm and chromatin. The most conservative position to take is that inherited epigenetic factors influencing development are limited to chromatin and cytoplasm. Some thinkers in the developmental systems theory tradition propose other factors, extracellular and even extrasomatic, as long as they reliably recur in succeeding generations (Griffiths and Gray 2001, 196), while Susan Oyama, closer in outlook to Deleuze, insists on the evolutionary significance of "the variability of individual ontogenies" (Oyama 2009, 150–51). But even if we stick to intracellular elements as the limits of our inherited epigenetic factors influencing development, we have to recognize that cell position in development plays a role in cell differenciation.[8]

Wherever one draws the line of inherited epigenetic factors influencing development—Jablonka and Lamb will go so far as to extrasomatic factors (e.g., a particular ambient temperature controlled by niche-construction) or more complex "scaffolding" operations, including exposure to language and other symbolic systems—we can see that even on conservative readings, gene regulation networks are, to borrow a term from chapter 4, dynamically interactional; they are dispersed, multifactorial, and no longer simply genomic. From this perspective, development is the key to seeing the limits of genetic determinism. We need not go as far as West-Eberhard (2003, 93) when she talks of the "enslavement of the genome by the phenotype," but we have to see the dethroning of DNA as master molecule (as localized and transcendent to the process) and its recruitment into a functional role in a distributed and differential system. To speak bio-ontologically, the genome is no longer transcendent but immanent; to speak biopolitically, it is no longer the monarch but now a team player; to speak Deleuzian, it is now part of dynamic interactionist networks that include epigenetic processes.

West-Eberhard's Eco-devo-evo Approach

We are now in a position to discuss West-Eberhard's (2003) *Developmental Plasticity and Evolution*. West-Eberhard calls her work "developmental evolutionary biology," which we can call "devo-evo," in contrast to the more well-known "evo-devo" (vii). In West-Eberhard's view, evo-devo has had a genetic focus, investigating how regulatory gene networks have evolved. The breakthrough toward evo-devo came when researchers found that key parts of the genome of many organisms are conserved over vast periods of time and shared by widely divergent phyla (Carroll [2005] is an excellent survey of evo-devo). A big discovery was homeotic genes, which structure development, acting as genetic switches controlling transcription factors regulating gene expression (turning them on and off). They are essential in body plans and are expressed in the order in which they are found in the chromosome; they control body segmentation, for instance. They are found in many different orders, conserved from before the division between arthropod and mammal. A famous one is *eyeless*. When transplanted from a mouse into a fly, it induces an eye formation. But there is a catch: the eye that forms is a fly eye, for it is the fly context that determines what kind of eye is formed (Carroll 2005, 67–68).

West-Eberhard asks, in seeking to broaden the molecular focus of evo-devo, what about the organism? After all, organisms are plastic—they can produce many different phenotypic expressions in response to different environments even though they have the same genetic makeup. In other words, development is flexible. But it is also robust: even with different genetic makeups, organisms still follow similar developmental pathways. So if phenotypically diverse organisms share genes, what is the source of their diversity? Her answer is that different developmental processes change the pattern of expression of the genes. But how do those different developmental processes become evolutionarily significant? West-Eberhard proposes that genetic control mechanisms can be exposed to selection by the phenotypic adaptation of organisms to new kinds of environment. For her, this phenotypic adaptation, what she calls *plasticity*, ultimately drives evolution.

West-Eberhard (2003, 20) proposes three key points here in linking development to evolution. First, she claims that environmental induction is a major initiator of evolutionary change; her slogan is that genes are followers, not leaders. Second, evolutionary novelties result from the reorganization of preexisting phenotypes and incorporation of environmental elements so that novel traits are not de novo results of mutation. Third, phenotypic plasticity can facilitate evolution by immediate accommodation and exaggeration of change; that is, phenotypic plasticity is not mere noise obscuring genetic patterns. From West-Eberhard's point of view, then, evolutionary biology must pay attention to "variation and selection within populations, speciation, developmental plasticity, and the origin of behavioral, physiological and life history traits" (vii). This means that while West-Eberhard does not deny natural selection, she focuses attention on the exposure to selection of gene expression networks developed by environmentally induced phenotypic variants. With her emphasis on environmental factors, we can add an "eco" and call her position "eco-devo-evo" (see also Gilbert 2001).

The resulting theory, although it invokes environmental influences on developmentally led evolution, is not Lamarckian, West-Eberhard emphasizes, because there is no direct influence of environment on genotype (West-Eberhard 2003, 192).[9] In other words, Lamarck thought that adaptive phenotypic changes were the source of variants that could be inherited (in contemporary terms, adaptive phenotypic changes produce genetic variation). But that is not West-Eberhard's scheme. What she says is that some adaptive phenotypic change is the result of developmental plasticity calling on previously hidden, that is, unexpressed, genetic variation (437). In other words, development does not produce genetic variation, but it calls on an untapped potential for the production of functional genes hidden in the genome.

The key concept for West-Eberhard here is "genetic accommodation" (147–57). The process goes like this: in exercising its plasticity, a new phenotype develops by being induced via a genetic mutation or an environmental difference. What has happened here in the latter case is that the new environment has brought forth an untapped potential of the preexisting genetic variation; such unexpressed genetic variation was previously hidden from selection by developmental robustness, that is, by the fact that there are many genetic pathways to the same phenotypic expression. It makes sense, given our previous discussion, that not all genetic variation is expressed, nor is all expression functional; remember that genetic expression and functionality depends on cellular and environmental conditions.

The key and controversial assumption is that this new phenotype is adaptive. The notion of adaptive phenotypic accommodation is called the "two-legged goat effect" (51–54) from the example of a goat born with two legs that changed many things in its phenotype to survive to reproductive age (though it did not, in fact, reproduce). The principle is that organisms can adaptively change in response to mutations or environmental changes and that these adaptive changes can become genetically accommodated. This change in phenotype creates new selection pressures because selection occurs through the interaction of phenotypes. The new phenotype starts to spread (as long as, in the case of environmentally induced change, the new environmental conditions reliably recur). Then the new selection pressures go to work on the new gene expression networks, which call on preexisting but unexpressed genetic variation; that is to say, the new gene expression networks are genuinely creative, putting together raw materials in a new combination. In this way, the new phenotype can eventually become a fixed expression (i.e., the gene expression networks can be selected for), even when the original environmental novelty is no longer present.

Now if the trait appears without recurrence of the environmental stimulus, there is *genetic assimilation*, whereas *genetic accommodation* is the general case in which the trait appears with or without the environmental stimulus. If it occurs only with the environmental stimulus, it is said to be an *environmentally sensitive* trait expression. In this latter case, what gets selected for is, conservatively speaking, the regulatory gene network or, radically speaking, the life cycle that includes the extended network encompassing the recurrent environmental stimulus and the regulatory gene network. But here is the key: the unexpressed genetic variation allowing for the production of new functional genes is only a potential to be actualized by the construction of gene expression networks. There are many unexpressed DNA strings available, but it takes the intensive (distributed and differential) dynamic system to create a novel functional gene by splicing and editing, excising introns and bringing together exons.

So, to recap, when an adaptive phenotypic change has a genetic component, the gene expression networks for this adaptive phenotypic variant will now be selected, but these gene expression networks are only potentials of the distributed and differential system working on preexisting but unexpressed genetic variation. So these gene networks are thus "followers" rather than "leaders" in evolution; in other words, plasticity in development precedes consolidation in heredity.

Deleuzian Development

I now turn to Deleuze to make the case that his notion of the virtual provides an ontological framework to enable us to understand the potentiality underlying West-Eberhard's notion of unexpressed genetic variation. As we know, Deleuze's perspective posits three interdependent ontological registers: (1) *actual* organisms with developed–differenciated organs, functions, and behaviors; (2) *intensive* developmental or individuating processes; and (3) *virtual* multiplicities or differentiated genetic–somatic–environmental networks.

Let me spell out the use of "virtual" here a bit. Recall the difference between functional genes as end products of the transcription process and hereditary genes as strings of DNA on the chromosomes. The constructed nature of functional genes implies that they are only potentially there in the hereditary genes; with Deleuze's help, we can see that such potentiality is virtual, that is to say, using Keller's metaphor, gene expression networks are activated in the musical performance in which epigenetic factors play—and thereby rewrite—the genomic score. In other words, we find distributed and differential systems of dynamic interaction between genomic and epigenetic processes. Furthermore, the differential nature of the elements, relations, and singularities of these distributed systems makes them a multiplicity in Deleuze's sense. That the *elements* are defined reciprocally means that they are networked together so that there is no such thing as a single, isolated "gene," or a single, isolated cell position, or a single, isolated niche; there are instead only networks of genes, cells, and niches, the functions of each element being dependent on the network state at any one time. That the *relations* are differential means that guiding development consists in the orchestration of linked rates of change of dynamically interactive processes of gene expression, protein folding, cell differenciation and migration, and so on, and that the relations contain *singularities* means that remarkable points in the relations of these processes serve as turning points or thresholds that induce starts and stops, or accelerations and decelerations, of the processes. We can thus propose virtuality as the ontological status of functional genes relative to hereditary genes. What shows up at the end of the transcription and translation process is not a mere tracing of a preexisting form; it is a genuinely creative process that integrates a differential field. In fact, as we will see, it is the intensive developmental processes that structure the virtual potentials; to return one last time to Keller's metaphor, "the problem is not only that the music inscribed in the score does not exist until it is played, but that the players rewrite the score (the mRNA transcript) in their very execution of it" (Keller 2000, 63).

Let us now turn to Deleuze's treatment of development in *Difference and Repetition.*[10] A very important point for Deleuze in his account of the biological model for ontogenesis is the priority of individuation to differenciation. In other words, singular differences in the genesis of individuals must precede the categories into which they are put; creative novelty must precede classification. As Deleuze puts it, "individuation precedes differenciation in principle.... Every differenciation presupposes a prior intense field of individuation. It is because of the action of the field of individuation that such and such differential relations and such and such distinctive points (preindividual fields) are actualized" (Deleuze 1994, 247). Individuation is thus the answer to the dramatic question "who?" rather than the essentialist question "what?" Individuals are singular events before they are members of species or genera; a species is a construct, an abstraction from a varying population of singulars. Deleuze insists that we beware the "tendency to believe individuation is a continuation of the determination of species" (247). Deleuze puts it very strongly: "any reduction of individuation to a limit or complication of differenciation compromises the whole of the philosophy of difference. This would be to commit an error, this time in the actual, analogous to that made in confusing the virtual with the possible" (247). The key point is that "individuation does not presuppose any differenciation; it provokes it" (247; translation modified). In other words, Deleuze must distinguish between any comparable difference between individuals—difference within a horizon of resemblance (i.e., representation), which can be classed in genus and species—and divergent difference or "individual difference," the "differenciation of difference," that which does not track genus and species but produces it via natural selection as a stabilizing procedure. Making species turn around individual and diverging difference is Darwin's "Copernican Revolution" (Deleuze 1994, 247–49; see also Deleuze and Guattari 1987, 48).

In seeking a concrete example of the precedence of individuation, Deleuze now turns to embryos, where he must finesse what looks to be a contradiction to his insistence on the priority of individuation to differenciation. Commenting on von Baër, Deleuze admits that embryonic life goes from more to less general.[11] However, this generality "has nothing to do with an abstract taxonomic concept" (i.e., it is not produced by differenciation as conditioning the comparison of the properties of finished products in a classification scheme) but is "lived by the embryo" in the process of individuation–dramatization (Deleuze 1994, 249; emphasis original). Recall that in "Introduction I," we discussed how the "order of reasons" from virtual through intensive to actual is "differentiation–individuation–dramatization–differenciation (organic and specific)" (Deleuze 1994, 251; 1968, 323; translation modified to restore the placement of the parentheses in the original). Thus the "experience" of the embryo (dramatization as spatiotemporal dynamism or morphogenetic process, the third element in the order of reasons) points backward, as it were, to the first two elements of the order of reasons (differentiation and individuation as field) rather than forward to the fourth element (differenciation). The experience of the embryo points to differential relations or virtuality "prior to the actualization of the species," and it points to "first movements" or the "condition" of actualization, that is, to individuation as it "finds its field of constitution in the egg" (Deleuze 1994, 249). This means that the lived generality of the embryo points "beyond species and genus" to the individual (i.e., to the field of individuation and that process of individuation) and to preindividual singularities rather than toward "impersonal abstraction" (249). So even though the specific form of the embryo appears early, this is due to the "speed and relative acceleration" of the elements of the individuation process, that is, to the "influence exercised by individuation upon actualization or the determination of the species." Thus a species is an "illusion—inevitable and well founded to be sure—in relation to the play of the individual and individuation" (249–50).

At this point, Deleuze provides a fascinating critique of genetic determinism. First, we are reminded again of the primacy of individuation over differenciation and that the "embryo is the individual as such caught up in the field of its individuation" (250). After the famously gnomic phrase "the world is an egg" (251), we read that "the nucleus and the genes designate only the differentiated matter—in other words, the differential relations which constitute the pre-individual field to be actualized; but their actualization is determined only by the cytoplasm, with its gradients and its fields of individuation" (251). By showing how genetic expression in ontogenesis is determined by cytoplasmic conditions, Deleuze is thus prefiguring the move in contemporary biology away from a self-identical and transcendent genetic program to a differential network of genetic and epigenetic factors controlling development.[12] This move to a differential virtual structuring organic individuation matches the Deleuzian principle of critique, the outlawing of the tracing relation between transcendental–virtual and empirical–actual, a principle that commands a nonresemblance of actualized species and parts to virtual differential relations and singularities. But the important point is that Deleuzian critique also commands the nonresemblance of both virtual multiplicity and actual adult individual to the intensive processes of morphogenesis or to what Deleuze calls the lived experience of the embryo. The twists and folds of embryogenesis do not resemble either the virtual network of relations among DNA strings and epigenetic factors or the actual structures and qualitatively different cell types of the adult organism.

To continue this all-too-brief sketch of organic individuation in *Difference and Repetition,* we see that for Deleuze, the "principal difficulty" of embryology is posing the field of individuation formally and generally (252). Eggs thus seem to depend on the species. But this reverses the order in which individuation precedes differenciation. So we must conceive individuating difference as individual difference: no two eggs are identical (252). Organic individuation, field and process together, is a singular event preceding differenciation. Once we have seen that, we have traced the order of reasons from (1) preindividual virtual differentiation through (2) the impersonal and intensive field of individuation to (3) spatiotemporal dynamisms as the process of individuation or dramatization to (4) differenciation as the formation of species and "parts," that is, qualitatively different cell types and functions that can then be classified into taxonomic schemes.

We have seen the relation of individuation and differenciation and the priority of the former. What about the relation between individuation and differentiation? Can changes in intensive processes change the virtual? Deleuze has to answer yes here to avoid the virtual being a Platonic realm of essences. Concerning the relation between individuation and differentiation, Deleuze writes in *Difference and Repetition* that "individuation is the act by which intensity determines the differential relations to be actualized, along the lines of differenciation and within the qualities and extensities it creates" (246; translation modified).[13] Writing a few pages later about intensities, Deleuze tells us that the expression of Ideas in intensities "introduces a new type of distinction into these [differential] relations and between Ideas a new type of distinction" (i.e., from the relation of virtual coexisting to relations of simultaneity or succession). He then writes that "all the intensities are implicated in one another, each in turn both enveloped and enveloping, such that each continues to express the changing totality of Ideas, the variable ensemble of differential relations." He concludes that "each intensity clearly expresses only certain relations or certain degrees of variation. Those that it expresses clearly are those on which it is focused when it has the *enveloping* role" (252; emphasis original).

I claim that the selective "focus" by which intensities clearly express only certain relations will itself introduce changes into the realm of Ideas. This means that countereffectuation is creative; experimentation in intensive morphogenetic processes will link together new combinations of differential relations, thereby forming new multiplicities. And this in turn will express or determine new potentials of the virtual.[14] That is what I take "determines the differential relations to be actualized" to mean: it renders them determinate in the sense of linking together previously unrelated relations. In pushing this interpretation, I want to avoid reading in Deleuze a Platonism in which the Ideas are already determined and so expression is mere copying of already made linkages of relations.

Countereffectuation and Genetic Accommodation

Deleuze insists on the primacy of individuation over differenciation. Differenciation is the classification of new species, whereas individuation is the production of new individuals. This is the equivalent of West-Eberhard's emphasis on developmental plasticity leading the way to evolutionary change. Recall how developmental plasticity is the creativity of the dynamic interaction of phenotype and environment. When an adaptive phenotypic change has a genetic component, the gene expression networks (or more radically, the life cycle) for this adaptive phenotypic variant will now be selected (if the environmental change reliably recurs). Now these accommodated or newly and creatively expressed gene expression networks (again, more radically put, the life cycle provoking the extended system of regulatory gene network and recurrent environmental conditions) were only "virtual," that is, potentials, only revealed ex post facto, of the preexisting but unexpressed genetic variation. To paraphrase Deleuze on Spinoza: we do not know what a body can do until it has done it.

Here we see the meaning of West-Eberhard's phrase that gene networks are thus "followers" as opposed to "leaders" in evolution. Instead of being the sole causal factors, they are often just "bookkeeping." That is, it is developmental plasticity and phenotypic adaptivity (in Deleuze's terms, intensive processes of individuation) that take the lead and bring out the potential to form the gene expression networks creating functional genes on the basis of the previously unexpressed potentials of hereditary DNA. But here is the crucially important point: the potentiality of the hereditary DNA is not preformationism: there is no present–actual–homuncular–already-determined "unit" or "program" in the DNA that determines the actualization of the potential. The virtual is not "self-determining": it is determined, on the spot, each time, by the individuation process.[15] It is the individuation process that takes the lead, which has to creatively produce something new in the world. This priority of individuation is what West-Eberhard talks about as developmental plasticity and phenotypic adaptivity and is a perfect example of the reality of creative countereffectuation, the anti-Platonic move in which experimental intensive process changes virtual structures.

Notes

Introduction I

1 Although it is an excellent book, Braver (2007) nonetheless treats only Foucault and Derrida, not Deleuze, in its "History of Continental Anti-realism."

2 Naturalism is a notoriously fecund notion. We can say that Deleuze is a naturalist qua antihumanist, in the Spinozist sense of denying that humans form a "kingdom within the kingdom" (of nature). In refusing a special status to human beings, Deleuze uses the same basic concepts (self-organization and creative novelty in dynamic systems) to handle phenomena in the physical, biological, social (including the social animals), and human registers. However, these same basic concepts have enough differences in their expression in the different registers that we cannot say Deleuze is a reductionist for whom purely physical explanations are the only form of legitimate discourse.

3 Not all transformation points in systems have singularities in their model, as this paragraph implies. The important point is that the structures of state spaces can suggest a different kind of causality than linear chains of efficient causality. We can use a nontechnical vocabulary to express this new conceptuality: patterns, thresholds, triggers, sensitive zones, etc. I will retain the singularity terminology, though, because (1) singularities do appear in the models of some systems and because (2) Deleuze uses the term, in a not always technically precise manner, to mean that which determines the internal structure of a state space, that is, the layout of the attractors. (I would like to thank Chuck Dyke and Alistair Welchman for advice on this issue.)

4 The most conservative explanation is that such systems are merely unpredictable rather than undetermined. In other words, *chance* is here just an epistemological term for the limits of our prediction; it is not ontologically grounded in the sense of allowing the system to escape from deterministic laws.

5 Notable discussions of *Difference and Repetition* include DeLanda (2002), Williams (2003), Beistegui (2004), Bell (2006), Toscano (2006), Bryant (2008), and Hughes (2009). There are complex technical discussions in the current scholarly literature as to the relation of the project of *Difference and Repetition* (Deleuze 1994) and of *A Thousand Plateaus* (Deleuze and Guattari 1987). Roughly speaking, the question is that of naturalization, with *A Thousand Plateaus* being the full naturalization of the somewhat still Kantian notion of a transcendental philosophy in which objects "incarnate" Ideas present in *Difference and Repetition.* Thus, instead of the notion of *vice-diction* in *Difference and Repetition* (Deleuze 1994, 189), whereby "thought" moves from an actual product to its transcendental conditions, we find the notion of *destratification* in *A Thousand Plateaus,* the move of systems themselves from a homogeneous and centralized "stratified" condition to a fully material heterogeneous and distributed "destratified" condition. This destratified condition allows systems to converge on a "plane of consistency" that allows the construction of novel assemblages. Ontologically speaking, then, in the move from *Difference and Repetition* to *A Thousand Plateaus,* intensive "spatiotemporal dynamisms" are no longer schemalike dramatizations mediating virtual Ideas and actual objects (Deleuze 1994, 218) but are simply immanent material haecceities, compositions of the movements and affects of bodies (Deleuze and Guattari 1987, 260–61). The disappearance of the transcendental or flattening of the ontology in *Difference and Repetition* and in the passage from *A Thousand Plateaus* is clearly explained by Toscano (2006, 175–80); there remains, however, a difficult question as to the status of the abstract machine(s) and "diagrams" in the latter work (Deleuze and Guattari 1987, 510–14; see Bonta and Protevi 2004, 47–49). I note the importance of Welchman (2009) in chapters 8 and 9 when broaching the question of panpsychism in light of the naturalization question.

6 Differential calculus for Deleuze is only a "technical model" for the structure of progressive determination of the Idea (Deleuze 1994, 220–21). The important thing is the ontological differences among virtual, intensive, and actual (terms we will discuss shortly in the main body of the text). This tripartite difference is illustrated by the difference between *differentiation* as determining the existence and distribution of singularities in a vector field and *integration* as the full determination of singularities while generating the trajectory modeling system behavior (Deleuze 1994, 176–79; DeLanda 2002, 30–34). But

the physical model of crystallization, the meteorological model of tropical cyclones, the biological model of gene regulatory networks and protein synthesis, and the social model of disciplinary institutions are also models of individuation leading the actualization of a virtual field. My thanks are due to James Williams for having insisted—more times than should have been necessary—on the importance of this point for my treatments of Deleuze and the sciences.

7   Deleuze distinguishes intensive processes—those that cannot change beyond a certain threshold without qualitative change of the behavior pattern of the process—from extensive properties, which can so change. In a simple example, a ruler cut in half becomes two rulers (length is thus an extensive property), whereas a pot of water heated from below produces convection currents at a certain threshold of temperature difference between top and bottom (heating water is thus an intensive process, or in standard terminology, temperature is a control parameter of the system) (DeLanda 2002, 26–27).

8   Deleuze's discussion of the "progressive determination" of an Idea is expressed in the language of calculus (Deleuze 1994, 171), but this is only for expository reasons. Following DeLanda's (2002, 30–31) discussion, we see that considered as pure "elements," rates of change are undetermined but determinable (dx, dy). As these rates of change are linked, they enter into differential relations, which are reciprocally determined (dx/dy)—this is differentiation as yielding instantaneous rates of change. These differential relations define the "existence and distribution" of the singularities of a vector field, but those singularities are only completely determined as those differential relations and singularities are actualized (values of dy/dx)—this is integration as the generation of trajectories.

9   If the context does not require distinguishing the field of individuation from the process of dramatization, I will sometimes simply write "intensive individuation process" to denote both aspects.

10 The main works here are Oyama (2000), Lewontin (2002), and Oyama, Griffiths, and Gray (2001). Developmental systems theory themes are also treated in Pigliucci and Müller (2010).

11 See Varela (1999b, 302) for the notion of "mutual bootstrapping" of trajectories and attractors in a dynamic system model, which we can gloss as the countereffectuating relation of intensive processes changing virtual structures. Countereffectuation will also change the conditions for the sense of past processes, but exploring this would

take us deep into the thickets of *Logic of Sense* (Deleuze 1990). Williams (2008) is an excellent guide here.

12   Concerning the relation between individuation and differentiation, Deleuze (1994, 246; translation modified) writes in *Difference and Repetition* that "individuation is the act by which intensity determines the differential relations to be actualized, along the lines of differenciation and within the qualities and extensities it creates." I discuss this point in chapter 10.

13   Deleuze distinguishes "dramatic" "who?" questions, which pick out individuation processes, from essentialist "what?" questions, which classify products. However, our focus on the individuation of a hurricane, such that it deserves a proper name, cannot be so extreme that we lose sight of the shared structure of the morphogenetic process leading to hurricanes. Each hurricane is unique—they really do deserve proper names—yet they are all hurricanes. In other words, Deleuze does not deny the utility of genera, but he does insist that individuation precedes the differenciation of genera; this insistence allows for countereffectuation and hence for dynamic development in the virtual register. These very delicate points are discussed in chapter 5 of Deleuze (1994, 244–54).

14   Deleuze rehearses in several works the debate between Geoffroy Saint-Hilaire and Cuvier, which follows the same lines of a distinction between the structure of morphogenetic processes and the classification of properties of products (Deleuze and Guattari 1987, 45–47; Deleuze 1994, 184–85). For commentary, see Ansell-Pearson (1999, 160) and Sauvagnargues (2004, 138–45).

15   My thanks to Manuel Cabrera Jr. for pointing me to this passage and for clarifying remarks.

16   Deleuze (1994, 126) insists on a differential reading of Nietzsche's eternal return: "the same is said of that which differs and remains different. The eternal return is the same of the different, the one of the multiple, the resemblant of the dissimilar."

Introduction II

1   This second section of the introduction is the only piece of the book that does not mention Deleuze by name, but the framework it develops concerning the multiple chronological and compositional registers of "bodies politic" is thoroughly Deleuzian in spirit and will serve to introduce the themes of the following chapters.

2   In "Not One, Not Two," Varela (1976, 63) also notes the synchronic emergence of wholes from the interaction of parts.

3   I am not claiming that *all* systems of organization closure, e.g., those that are in Varela's terms informational, are dangerous social models. Thus I'm not arguing that Varela's warning holds against all social systems, which Luhmann terms *autopoietic.*

4   Advances in computer simulation will later allow for the dynamic modeling of cooperation and competition in the formation of resonant cell assemblies, as we can see in *The Embodied Mind* (Varela, Thompson, and Rosch 1991).

5   Robert Rosen (1991) takes up category theory; Varela drops the formalization as more adequate dynamical models appear.

6   In a dialogue with Cornelius Castoriadis, Varela specifies that such emergence is neither aleatory nor calculable (Castoriadis 2000, 113).

7   This approach has been modified and developed by, among others, Thompson (2007), Clark (1997), and Noë (2004).

8   Iris Marion Young's (1990; 2005) "Throwing Like a Girl" is a classic critique of the privileged and empowered masculine corporeal subject presupposed in Merleau-Ponty's analyses. She shows how many feminized corporeal subjects experience parts of the world as anxiety-producing obstacles, the "same" parts that a competent masculinized subject will encounter as amusing occasions for the demonstration of competence.

9   Regarding the emergent functions in the interplay of animal physiology and the structures they build, see Turner (2000). Regarding human–technology interfaces, see Clark (2003) and Hansen (2006).

10   See also "At the Source of Time" (Varela and Depraz 2005, 75), in which we read of "a primordial duality, a rough topology of *self-other.*"

11   Rogers and Hammerstein, in *South Pacific:* "you've got to be taught . . . carefully taught!"

12   This is where we need more empirical work on humans and mirror neurons. With monkeys, we know that it is simply intraspecific. What I want to ask is if that is the case for humans or if our historical–cultural bodily formation (what I am calling *political physiology*) doesn't set even our mirror neuron empathizing at socially constructed mid-level categories. Do we dehumanize enemies in warfare (we can kill them because they are inhuman vermin, insects, rats, etc.) or is simple racialization enough? Is it that the inferior races are humanly liminal, at the border of animals?

1. Geo-hydro-solar-bio-techno-politics

1   In translating *Mille Plateaux* (Deleuze and Guattari 1980) into English, Brian Massumi uses two English words to translate the French *terre*, which can mean both "earth" in the astronomical sense of our planet and "land" in the geographical sense of a cultivated area. As far as I have been able to determine, there is no consistency in Deleuze and Guattari's use of the majuscule in the French text; both *Terre* and *terre* are used in the sense of "earth" and "land." The Anglophone reader should keep in mind the close proximity of *terre* with *territoire* (territory).

2   There is a vast literature on Aristotle's notion of courage, and much of it includes commentary on Aristotle's own use of Homer in the *Nicomachean Ethics*. A recent work in the field is Ward (2001).

2. The Act of Killing in Contemporary Warfare

1   As the anonymous reviewer of the original version of this work reminded me, the standard evolutionary explanation for the adoption of signaling rather than fighting among nonhuman animal conspecifics does not involve empathetic identification but rather an instinctually embedded cost–benefit analysis. For example, the risk of harm from a fight outweighs the benefits of mating so that it is better to accept defeat and wait to find another opportunity later.

2   E.g., "the physical experience of being underneath a wave of water seems to be secondary to the psychological experience. The person's mind believes he is drowning, and his gag reflex kicks in as if he were choking on all that water falling on his face" (Layton 2011).

3   Niehoff (1999) offers support here. See p. 75 on "protective aggression," citing Archer (1988); p. 127 on the release of norepinephrine in attack situations; and p. 130 for a summary of Gray (1977), which postulates a behavioral inhibition system tied to physiological arousal.

4   Here we use computer metaphors, but we have to be very careful not to let them imply any stance on cognition as computation of discrete symbols. Wetherell (2012) rightfully insists that in the vast majority of cases, cortical control is quickly exerted. I accept this and would say that the "pure" state of direct linkage of the social and the somatic is a limit that is closely approximated in rare cases of berserker rage or frenzied panic.

5   See LeDoux (1996, 200–3) for a brief overview of emotional memory; although LeDoux focuses on fearful memories, dopamine would seem

to be a key player in the production of pleasant memories, as summarized by Niehoff (1999, 131).

6 This chapter would not have reached this stage without Roger Pippin (Department of Communication Studies, University of South Florida), who was coauthor of the first draft. As the essay has developed, however, we have decided it would be best for Protevi to assume authorship of this article and for us to pursue our joint project in another publication. Further thanks are due to Jane Richardson and, for comments and questions, to many attendees of the Cognition: Embodied, Embedded, Enactive, Extended conference organized by Shaun Gallagher at the University of Central Florida, October 20–24, 2007.

3. Music and Ancient Warfare

1 It would be very interesting one day to put Deleuze and Guattari's notion of affective categorization in connection with Elizabeth Rosch's (1978) prototype theory. Both challenge the Aristotelian notion of categories based on a set of necessary and sufficient conditions defining an essence. What we would need to do is to define a Deleuzoguattarian pedagogy that enables us to see the world in terms of affects, i.e., to see the world as a theater for transcendental empiricism: what can bodies do? We have to experiment!

2 As I note in chapter 2, Wetherell (2012) rightfully insists that in the vast majority of cases, cortical control is quickly exerted. I accept this and would say that the "pure" state of direct linkage of the social and the somatic is a limit that is closely approximated in rare cases of berserker rage or frenzied panic.

3 Cultural evolution is tied in with debates surrounding sociobiology and evolutionary psychology. An informative exchange can be examined in Fracchia and Lewontin (1999; 2005) and Runciman (2005a; 2005b). See also Lewontin (2005), commenting on Richerson and Boyd (2005).

4 For background on the interplay of genes and experience in neural development, see Wexler (2006) and Mareschal et al. (2007).

5 Damasio's somatic marker theory, and more generally, his somatic theory of emotion, are not without critics, who reject the somatic theory as based in what they see as an outmoded James–Lange tradition. Even among those sympathetic to the somatic theory, controversies remain concerning the precise role of cortical versus midbrain and brain stem structures in generating basic emotions. Some of the

debates within the field are accessibly summarized in Watt (2000) and Panksepp (2003) (on Damasio [1999; 2003], respectively).

6   For a brief review of the literature from the antiuniversalist position, see Sponsel (2000); for a book-length statement of the universalist position, see Otterbein (2004). For a brief review of the controversy over Steven Pinker's position, see Lawler (2012a). The problem with inferring inevitability in the future from universality in the past is tackled by Fry (2012), who not only shows peaceful societies but also isolates their characteristics that could be institutionalized today.

7   Bowles (2012, 877) has the correct argument with regard to warfare as a selection pressure on biocultural evolution: "the degree of mortal conflict and extent of genetic differences among ancestral forager groups were jointly sufficient to have allowed the evolution of a genetically transmitted predisposition to contribute to common projects (including defense and predation vis-à-vis other groups, even when one's individual fitness would have been enhanced by free riding on those who would 'aid and defend each other.' Whatever the balance of cultural and genetic factors in the evolution of human cooperativeness, between-group conflict almost certainly played a pivotal role."

8   For more on affect produced in collective religious rituals, see Lord (2007) and Ehrenreich (2007).

9   Diamond (1992) discusses waterborne parasites weakening the peasant population in irrigation regimes—an underexplored area of geo-hydro-bio-politics.

10   Although it came to my attention too late for full incorporation into the text, a brief glance at the notion of *forward panic* in Randall Collins's (2008, 83–132) *Violence: A Micro-sociological Theory* shows that it might be useful here.

4. Dynamic Interactionism

1   Dreyfus (1972; 1992) influenced the embodied mind school's critique of computationalism and connectionism. Technically speaking, however, Dreyfus's approach is better seen as Heideggerian than as belonging to the embodied mind school. That is to say, he rejects the notion of mind entirely so that it being embodied is not an advance for him. See Dreyfus (2007) for this argument. Among the major works of the embodied mind school proper are Varela, Thompson, and Rosch (1991), Clark (1997), Gallagher (2005), Noë (2004), Wheeler (2005), and Thompson (2007).

2 See Wheeler (2005) for a treatment of "action-oriented" representation. A radical antirepresentationalist approach is found in Chemero (2009).

3 The term *metastable* comes to Deleuze from Simondon (1995), but its independent use in Kelso (1995) provides evidence that Deleuze's ontology can inform our readings of dynamic systems research and, in particular, neurodynamics. On Simondon's "transindividuality," with reference to Spinoza, see Balibar (1993) and Sharp (2011, 34–37).

4 On the closely related "interactive brain hypothesis," see Di Paolo and De Jaegher (2012).

5. The Political Economy of Consciousness

1 I am not suggesting any pathological "thought insertion" here, merely the everyday phenomenon of being prompted to form a new thought after discussion with other people. Thought insertion, we could say, is an extensive phenomenon in which a fully formed thought is inserted into a personal mental sphere, whereas prompting of thought is an intensive individuation process.

2 See Vogel (2000) for a review of the classic Marxist feminist literature on domestic labor; the parallel is that hidden female domestic labor allows for the presentation of public male (supposedly self-constituted) identity.

3 A recent sociological survey with a long-term historical scope notes that "violence is ubiquitous in right-wing movements as an action and/or a goal. Violence can be strategic, chosen among alternative tactical actions to achieve a goal, often by highly insular groups intently focused on their perceived enemies. . . . Strategic violence is targeted at enemy groups, such as Jews, racial minorities, or federal government installations. Other right-wing violence is more performative. Performative violence binds together its practitioners in a common identity, as when white power skinheads enact bloody clashes with other skinhead groups and each other" (Blee and Creasap 2010, 276).

Meanwhile, a widely noted U.S. Department of Homeland Security (2009, 2) report states that "rightwing extremists have capitalized on the election of the first African-American president, and are focusing their efforts to recruit new members, mobilize existing supporters, and broaden their scope to propaganda, but they have not yet turned to attack planning." The department notes in particular that "a recent example of the potential violence associated with a rise in

rightwing extremism may be found in the shooting deaths of three
police officers in Pittsburgh, Pennsylvania, on 4 April 2009. The alleged gunman's reaction reportedly was influenced by his racist ideology and belief in antigovernment conspiracy theories related to gun
confiscations, citizen detention camps, and a Jewish-controlled 'one
world government'" (3).

4   When I was unemployed, some fifteen years ago, for six months, I was
often overcome with shame, no matter how often I reminded myself
of the objective factors, the nonsensical nature of the affect, etc. But
where did I pick up this shame? I cannot see how it was transmitted to me by another actual instance of shame. You could say I had
been socialized so that I carried a latent disposition to shame that
became occurrent in the right circumstances. But that is hardly less
"metaphysical" than an account of virtual or environmental collective
affective with shamed selves crystallized out of that. I do not think
we'll escape metaphysics that easily; there is a lot of potential versus
actual metaphysics to be worked out there in the latent–occurrent
disposition scheme, as I try to do in chapter 7.

5   Another topic for analysis would be the bike generators being set up
at OWS. In another possible blunder, recalling that of the banning
of bullhorns, the city confiscated gasoline generators prior to the
late October snowstorm. The brilliant OWS response was to acquire
bicycle generators. Will there be an analogous affective supplement
from taking turns on the bikes to generate electricity?

6   Faces are an extremely important factor in political affect. In analyzing OWS, we'd have to consider the use of the Guy Fawkes "V for
Vendetta" masks, the denunciation of "faceless corporations," and
the "faciality machine" in Deleuze and Guattari (1987).

6. The Granularity Problem

1   Feminist analyses of neuroscience are now appearing with greater
frequency (Fine 2010; Jordan-Young 2010; Bluhm, Jacobson, and Maibom 2012).

7. Adding Deleuze to the Mix

1   The latest 4EA work to tackle the realism–idealism debate is Chemero
(2009), which contains a fine overview of the issue (183–205).

2   I will use organismic behavior guided by sensorimotor perception as
my paradigm case. Among the differences that divide thinkers in the

4EA approach is the status of "representation-hungry problems" and "action-oriented representations," which are approved of by Clark (1997) and Wheeler (2005) but contested by Chemero (2009). These authors do agree, however, that a wide range of behaviors that traditional artificial intelligence or orthodox cognitive science tries to handle by positing computation performed on representation do not in fact require representations as they can be handled by models of coordination within the organism–environment couple considered as a dynamic system.

3 Despite the general trend described here, how precisely to define cognition within this general framework, and the exact nature of which biological details are necessary for cognition—or even if they are necessary at all—is hotly contested within the 4EA and orthodox approaches. See Wheeler (2010) for an overview of the issues involved.

4 *Neurological correlates* is a loaded term in this context and should be approached in terms that Chemero (2009, 200; emphasis original) lays out clearly: "Experiences do not happen in brains. Even though it is perfectly obvious that *something* has to be happening in neurons every time an animal has an experience, for the radical embodied cognitive scientist, as for the enactivist, this something is neither identical to, nor necessary and sufficient for, the experience."

5 A more full treatment of this issue would take us to the distributionist vs. localist dispute in neuroscience. Deleuze is on the side of the distributionists. Thus he would agree that, e.g., though the hippocampus may indeed be necessarily involved in long-term memory, the retrieval of a memory involves the integration of distributed neural systems. In many ways, the dispute between distributionists and localists is a dispute between dynamicists and anatomists, and Deleuze, as a process philosopher, will side with the dynamicists.

6 See Varela (1999b, 302) for the notion of *mutual bootstrapping* of trajectories and attractors in a dynamic system model, which we can gloss as the countereffectuating relation of intensive processes changing virtual structures.

7 Deleuze follows Bergson's critique of the possible as the retrojection of the real minus existence (Deleuze 1991, 43, 96–97).

8 Although in his books, Prinz (2004) does not rely on dynamic systems theory or on phenomenology, he does rely on biologically plausible models of emotion that emphasize the brain–body–world context: "emotions are not merely perceptions of the body but also

perceptions of our relations to the world.... This book... is an attempt to bring body, mind, and world together" (20).

9   Certainly learning to swim in the ocean is more complex than learning in a pool; but even in a pool, putting on one of the new bodysuits will require that even expert swimmers attempt a new "conjugation."

10   We should note that in his discussion of Affordances 1.1, Chemero insists that abilities are not dispositions, which, in Chemero's understanding, are automatically triggered under the right circumstances (145). Thus Chemero will claim that abilities are not inherent in animals (as are dispositions) but in animal–environment systems. However, I do not believe that Prinz's notion of disposition, discussed earlier, is as deterministic as Chemero's.

11   Think of two trains moving on parallel tracks at the same velocity. You can look from one window to the other and see stable things, but this is only due to the coordination of rates of change.

12   I would like to acknowledge very helpful comments from Jeff Bell, Manuel Cabrera Jr., Manuel DeLanda, Shaun Gallagher, Joe Hughes, Mike Wheeler, James Williams, and two anonymous reviewers from *Phenomenology and the Cognitive Sciences.*

8. Larval Subjects, Enaction, and *E. coli* Chemotaxis

1   The major commenters on *Difference and Repetition*—Hughes (2009), Bryant (2008), Beistegui (2004), and Williams (2003)—do not isolate the level of organic synthesis. The exception is Ansell-Pearson (1999).

2   For a treatment of the infinite regress problem in philosophical psychology, see Zahavi (2005).

3   Of course, Aristotle himself thought that plants possessed only the nutritive or vegetative psyche and that only animals had a sensible psyche. For an interesting take on the Aristotelian resonances here in the context of contemporary philosophy of mind and cognitive science, see Wheeler (1997).

4   We cannot treat the very rich discussion of the double aspect of death in *Difference and Repetition,* but we are here alluding to the way Deleuze reads the "death instinct" as "an internal power which frees the individuating elements from the form of the I or the matter of the self in which they are imprisoned . . . the liberation and swarming of little differences in intensity" (Deleuze 1968, 333; 1994, 259).

5   There is an archaic sense of the English word *sense* meaning "direction," as in "the sense of the river." This sense is still present in

French, as in, among other uses, the expression *sens unique* for "one-way street." I have treated the threefold "sense of sense" in Protevi (1990; 1998).

6 For contemporary critiques of genetic determinism, see the developmental systems theory school of thought, whose founding document is Oyama ([1985] 2000); see also Oyama, Griffiths, and Gray (2001).

9. Mind in Life, Mind in Process

1 This is a contested reading, but against Wheeler (2011), I read Thompson (2007) as upholding a coextensivity thesis regarding the relation of mind and life, rather than Wheeler's enrichment thesis, which would move from life to mind. I cannot fully engage with Wheeler's rich reading, but a key quote for me in defending the coextensivity thesis is the following: "any living system is both an autopoietic and a cognitive system. . . . This thesis is sufficient to establish a deep continuity of life and mind" (Thompson 2007, 127). In other terms, autopoietic (cellular or multicellular) life is sufficient for cognition; where there is such life, there is cognition. Leaving aside the ALife question and the ancient hylozoism question, which ask about noncellular life, the panpsychism question will ask about nonliving mind. Panpsychism asks whether autopoietic (cellular or multicellular) life is necessary for cognition (mind) or whether there is a defensible notion of mind not just in life but in process. Is mind a genus of which enactive cognition or mind in life is a species?

2 I will provide my own translation of the Simondon passages.

3 Metastability is well known in dynamic systems theory, serving, e.g., as a key term in Kelso (1995).

4 "The simplest organism, which we can call 'elementary,' is that which does not possess a medial interior milieu, but only an absolute interior and exterior" (Simondon 1995, 225).

5 Margulis's notion of symbiogenesis (Margulis and Sagan 1995) is echoed by Deleuze and Guattari (1987, 238). See Ansell-Pearson (1999, 165–66) for a brief discussion.

6 As we will see in chapter 10, West-Eberhard (2003) does not deny natural selection but claims it will favor the spread of a particular environmentally induced phenotypic variant when it has positive effects on individual fitness, i.e., when it is adaptive. West-Eberhard denies that this is Lamarckism because there is no direct influence of environment on genotype.

7   To recall a point we made in "Introduction I" and to which we will
    return in chapter 10, let us note the following key passage in *Difference
    and Repetition*: "individuation is the act by which intensity determines
    the differential relations to be actualized, along the lines of differencia-
    tion and within the qualities and extensities it creates" (Deleuze 1994,
    246; translation modified).

8   We can compare this to Donn Welton's notion of "transcendental
    space" of constitutive phenomenology and "transcendental time" of
    genetic phenomenology: "This gives us yet another interesting way
    of understanding the difference between constitutive and genetic
    analysis. We can say that constitutive phenomenology schematizes
    the structural transformations making phenomenal fields possible
    according to transcendental *space*. They are framed as layers or strata
    beneath each field, providing it with its supporting ground. Genetic
    phenomenology schematizes those transformations in terms of tran-
    scendental *time,* and thus as a process of development in which the
    earlier gives rise to the later, and in which the later draws and gives
    direction to the now" (Welton 2003, 254; emphasis original).

9   As Skrbina (2005, 196–98) notes, Bateson later backs away from this
    cybernetic mind position.

10. The Virtual Status of "Unexpressed Genetic Variation"

1   My thanks to Dennis Des Chene and Chuck Dyke for help with this
    chapter. I will not discuss Deleuze's relation to the biological think-
    ers whom he cites, but this is an important field of research already
    well under way. See Ansell-Pearson (1999) on Darwin; Bogue (2003)
    on Raymond Ruyer; Sauvagnargues (2004, 2005, 2009) on Gilbert
    Simondon, Georges Canguilhem, and Geoffroy Saint-Hilaire; and
    May (2005) and Marks (2006) on François Jacob and Jacques Monod.

2   "Unexpressed genetic variation" also appears in Badyaev (2005) and
    in Young and Badyaev (2007). The related notion of "ectopic expres-
    sion" (gene expression with no discernible function) is covered in
    Rodriguez-Trelles, Tarrio, and Ayala (2005). See also Allen (1997, 54)
    on the presence of a "pool of hidden adaptability" in evolutionary
    systems.

3   I take off from some thought-provoking lines in Evelyn Fox Keller's
    (2000) *The Century of the Gene*; I will define the technical terms in the
    body of the chapter. Keller writes that we might "consider the mature
    mRNA transcript formed after editing and splicing to be the 'true'

gene. But if we take this option (as molecular biologists often do), a different problem arises, for such genes exist in the newly formed zygote only as possibilities, designated only after the fact. A musical analogy might be helpful here: the problem is not only that the music inscribed in the score does not exist until it is played, but that the players rewrite the score (the mRNA transcript) in their very execution of it" (63).

4 I am aiming to present a model of the received or mainstream view that will unfortunately seem to be a straw man for practicing biologists, who will be able to cite the many nuances and hedges that were built into this view but that were overridden in popular presentation. Nonetheless, as the standard popular view in mainstream media and in the minds of most educated but not specialized people continues to have its effects, it is worth examining it here.

5 See also Doyle (1997). Keller's recent proposal for reformulating evolutionary and developmental language can be found in Keller (2010).

6 Note that unexpressed DNA can evolve by drift (mutation) and other processes, independent of natural selection. This is because natural selection only works on phenotypes. (Natural selection is about real-world interactions, even if they can be tracked by gene shifts.) The variation in unexpressed DNA can be a reservoir for genetic bases for developmental plasticity, as we will see in discussing West-Eberhard.

7 The proper biological term is *cellular differentiation,* but for reasons that should be clear by this point in the book, I am using the Deleuzian *differenciation* to denote this process, in which modifications of multipotential stem cells come to rest in different fixed cell types.

8 The classic article on positional information is Wolpert (1969); see also Wolpert (1989) for a twenty-year retrospective of the idea's origin and reception.

9 Jablonka and Lamb (2005, 107), it should be noted, owing to their emphasis on epigenetic inheritance, happily accept the term *Lamarckian.*

10 We should recall Deleuze's warning in *Difference and Repetition* that biology is only a "technical model" of the way intensive processes of individuation lead the way in the actualization of virtual multiplicities (Deleuze 1994, 220–21). As we saw in "Introduction I," the physical model of crystallization, the meteorological model of tropical cyclones, and the social model of disciplinary institutions are also examples of intensive individuation processes leading the actualization of a virtual field.

11 Although it does not appear in the bibliography of *Difference and*

Repetition, Deleuze and Guattari do refer to the original version of Canguilhem et al. ([1962] 2003) in *A Thousand Plateaus* (Deleuze and Guattari 1987, 522n9). It might be the source for the discussion of von Baër in *Difference and Repetition.*

12 I am using *prefigure* here because, despite the tantalizing potentials of this brief passage, it is still quite possibly oriented to the operon–regulatory gene–genetic program model of Jacob and Monod, which, as we have seen in our recapitulation of Keller's view, privileges genomic control by relegating epigenetic conditions to mere occasioning factors for regulatory gene action. See Marks (2006) for a treatment of Deleuze and Guattari's exploitation of the "intensive potentials" they find in the later works of Jacob and Monod.

13 The French reads, "L'individuation, c'est l'acte de l'intensité qui détermine les rapports différentiels à s'actualiser, d'après des lignes de différenciation, dans les qualités et les étendues qu'elle crée" (Deleuze 1968, 317).

14 See Varela (1999b, 302) for the notion of mutual bootstrapping of trajectories and attractors in a dynamic system model, which we can gloss as the countereffectuating relation of intensive processes changing virtual structures.

15 For a critique of the "sufficiency of the virtual," see Toscano (2006, 175–80).

Bibliography

Allen, Peter M. 1997. "Models of Creativity: Towards a New Science of History." In *Time, Process, and Structured Transformation in Archaeology*, ed. S. E. Van Der Leeuw and James McGlade, 39–56. London: Routledge.

Ansell-Pearson, Keith. 1999. *Germinal Life: The Difference and Repetition of Deleuze*. London: Routledge.

Archer, John. 1988. *The Behavioral Biology of Aggression*. Cambridge: Cambridge University Press.

Arquilla, John, and David Rondfeldt. 2000. *Swarming and the Future of Conflict*. Santa Monica, Calif.: RAND Corporation.

Averill, James R. 1982. *Anger and Aggression: An Essay on Emotion*. New York: Springer.

Badyaev, Alexander V. 2005. "Stress-Induced Variation in Evolution: From Behavioural Plasticity to Genetic Assimilation." *Proceedings of the Royal Society, B* 272:877–86.

Balibar, Etienne. 1993. "Spinoza: From Individuality to Transindividuality." http://www.ciepfc.fr/spip.php?article236.

Balter, Michael. 1998. "Why Settle Down? The Mystery of Communities." *Science* 282, no. 5393: 1442–45.

Barnes, Jonathan, ed. 1984. *The Complete Works of Aristotle*. Princeton, N.J.: Princeton University Press.

Basile, Pierfrancesco. 2009. "Back to Whitehead? Galen Strawson and the Rediscovery of Panpsychism." In *Mind That Abides: Panpsychism in the New Millennium*, ed. David Skrbina, 179–99. Amsterdam: John Benjamins.

Bataille, Georges. 1991. *The Accursed Share*. Vol. 1, *Consumption*. Trans. Robert Hurley. Cambridge, Mass.: MIT Press.

Bateson, Gregory. 1972. *Steps to an Ecology of Mind: Collected Essays in Anthropology, Psychiatry, Evolution, and Epistemology*. Chicago: University of Chicago Press.

Beistegui, Miguel de. 2004. *Truth and Genesis: Philosophy as Differential Ontology*. Bloomington: Indiana University Press.

Bell, Jeffrey A. 2006. *Philosophy at the Edge of Chaos: Gilles Deleuze and the Philosophy of Difference*. Toronto, Ont.: University of Toronto Press.

Berg, Howard. 2004. *E. coli in Motion*. New York: Springer.

Bergson, Henri. 1991. *Matter and Memory*. Trans. N. M. Paul and W. S. Palmer. New York: Zone Books.

———. (1913) 2001. *Time and Free Will: An Essay on the Immediate Data of Consciousness*. Trans. F. L. Pogson. Mineola, N.Y.: Dover.

Bispham, John. 2004. "Bridging the Gaps—Music as Biocultural Phenomenon." *ESEM Counterpoint* 1:78–81.

———. 2006. "Rhythm in Music: What Is It? Who Has It? And Why?" *Music Perception* 24, no. 2: 125–34.

Blee, Kathleen, and Kimberly Creasap. 2010. "Conservative and Right-Wing Movements." *Annual Review of Sociology* 36:269–86.

Bluhm, Robyn, Anne Jaap Jacobson, and Heidi Maibom. 2012. *Neurofeminism: Issues at the Intersection of Feminist Theory and Cognitive Science*. Basingstoke, U.K.: Palgrave Macmillan.

Boden, Margaret. 2006. "Of Islands and Interactions." *Journal of Consciousness Studies* 13, no. 5: 53–63.

Bogue, Ronald. 2003. *Deleuze on Music, Painting, and the Arts*. New York: Routledge.

Bonta, Mark, and John Protevi. 2004. *Deleuze and Geophilosophy*. Edinburgh, U.K.: Edinburgh University Press.

Bordo, Susan. 1986. "Anorexia Nervosa: Psychopathology as the Crystallization of Culture." *Philosophical Forum* 17 (Winter): 73–103.

———. 1993. *Unbearable Weight: Feminism, Western Culture, and the Body*. Berkeley: University of California Press.

Bowles, Samuel. 2012. "Warriors, Levelers, and the Role of Conflict in Human Social Evolution." *Science* 336 (May 18): 876–79.

Bowles, Samuel, and Herbert Gintis. 2003. "The Origins of Human Cooperation." In *The Genetic and Cultural Origins of Cooperation*, ed. Peter Hammerstein, 429–44. Cambridge, Mass.: MIT Press.

Braver, Lee. 2007. *A Thing of This World: A History of Continental Anti-realism*. Evanston, Ill.: Northwestern University Press.

Bray, Dennis. 2009. *Wetware: A Computer in Every Living Cell*. New Haven, Conn.: Yale University Press.

Bryant, Levi. 2008. *Difference and Givenness: Deleuze's Transcendental Empiricism and the Ontology of Immanence*. Evanston, Ill.: Northwestern University Press.

Burke, Carol. 2004. *Camp All-American, Hanoi Jane, and the High-and-Tight: Gender, Folklore, and Changing Military Culture*. Boston: Beacon Press.

Butler, Judith. 1989. "Sexual Ideology and Phenomenological Description: A Feminist Critique of Merleau-Ponty's *Phenomenology of Perception*." In *The Thinking Muse: Feminism and Modern French Philosophy*, ed. Jeffner Allen and Iris Marion Young, 85–99. Bloomington: Indiana University Press.

———. 2011a. "Bodies in Alliance and the Politics of the Street." http://www.eipcp.net/transversal/1011/butler/en.

———. 2011b. "Judith Butler at Occupy Wall Street." http://www.youtube. com/watch?v=JVpoOdz1AKQ.

Butzer, Karl. 1976. *Early Hydraulic Civilization in Egypt: A Study in Cultural Ecology.* Chicago: University of Chicago Press.

Canguilhem, Georges, Georges Lapassade, Jacques Piquemal, and Jacques Ulmann. (1962) 2003. *Du développement à l'évolution au XIXᵉ siècle.* Paris: Presses Universitaires de France.

Carroll, Sean B. 2005. *Endless Forms Most Beautiful: The New Science of Evo-Devo.* New York: W. W. Norton.

Casey, Edward. 2000. *Remembering: A Phenomenological Study.* Bloomington: Indiana University Press.

Castoriadis, Cornelius. 2000. *Post-Scriptum sur l'insignifiance, suivi de Dialogue.* Paris: l'Aube.

Chemero, Anthony. 2009. *Radical Embodied Cognitive Science.* Cambridge, Mass.: MIT Press.

Clark, Andy. 1997. *Being There: Putting Brain, Body, and World Together Again.* Cambridge, Mass.: MIT Press.

———. 1999. "An Embodied Cognitive Science?" *Trends in Cognitive Sciences* 3, no. 9: 345–51.

———. 2003. *Natural-Born Cyborgs: Minds, Technologies, and the Future of Human Intelligence.* New York: Oxford University Press.

Clark, Andy, and David Chalmers. 1998. "The Extended Mind." *Analysis* 58, no. 1: 7–19.

Clastres, Pierre. 1989. *Society against the State: Essays in Political Anthropology.* Trans. Robert Hurley, in collaboration with Abe Stein. New York: Zone Books.

Collins, Randall. 2008. *Violence: A Micro-sociological Theory.* Princeton, N.J.: Princeton University Press.

Correll, Joshua, Geoffrey R. Urland, and Tiffany A. Ito. 2006. "Event-Related Potentials and the Decision to Shoot: The Role of Threat Perception and Cognitive Control." *Journal of Experimental Social Psychology* 42:120–28.

Crisafi, Anthony, and Shaun Gallagher. 2009. "Hegel and the Extended Mind." *Artificial Intelligence and Society* 25, no. 1: 123–29.

Cross, Ian. 2003. "Music and Biocultural Evolution." In *The Cultural Study of Music: A Critical Introduction,* ed. Martin Clayton, Trevor Hebert, and Richard Middleton, 10–19. London: Routledge.

Damasio, Antonio. 1994. *Descartes' Error.* New York: Avon.

———. 1999. *The Feeling of What Happens.* New York: Harcourt.

———. 2003. *Looking for Spinoza.* New York: Harcourt.

Davies, John. 2005. "Linear and Nonlinear Flow Models for Ancient Economics." In *The Ancient Economy: Evidence and Models,* ed. J. G. Manning and Ian Morris, 127–56. Stanford, Calif.: Stanford University Press.

Dawson, Doyne. 1999. "Evolutionary Theory and Group Selection: The

Question of Warfare." *History and Theory* 38, no. 4: 79–100.

De Jaegher, Hanne, and Ezequiel Di Paolo. 2007. "Participatory Sense-Making: An Enactive Approach to Social Cognition." *Phenomenology and the Cognitive Sciences* 6, no. 4: 485–507.

De Jaegher, Hanne, and Tom Froese. 2009. "On the Role of Social Interaction in Individual Agency." *Adaptive Behavior* 17:444–60.

De Jaegher, Hanne, Ezequiel Di Paolo, and Shaun Gallagher. 2010. "Does Social Interaction Constitute Social Cognition?" *Trends in Cognitive Sciences* 14, no. 10: 441–47.

DeLanda, Manuel. 2002. *Intensive Science and Virtual Philosophy*. London: Continuum.

Deleuze, Gilles. 1968. *Différence et répétition*. Paris: Presses Universitaires de France.

———. 1988. *Spinoza: Practical Philosophy*. Trans. Robert Hurley. San Francisco: City Lights.

———. 1990. *Logic of Sense*. Trans. Mark Lester, with Charles Stivale. New York: Columbia University Press.

———. 1991. *Bergsonism*. Trans. Hugh Tomlinson and Barbara Habberjam. New York: Zone Books.

———. 1993. *The Fold: Leibniz and the Baroque*. Trans. Tom Conley. Minneapolis: University of Minnesota Press.

———. 1994. *Difference and Repetition*. Trans. Paul Patton. New York: Columbia University Press.

———. 2004. "The Method of Dramatization," trans. Michael Taormina. In *Desert Islands and Other Texts*, ed. David Lapoujade, 94–116. New York: Semiotext(e).

Deleuze, Gilles, and Félix Guattari. 1980. *Mille Plateaux*. Paris: Minuit.

———. 1984. *Anti-Oedipus*. Trans. Robert Hurley, Mark Seem, and Helen R. Lane. Minneapolis: University of Minnesota Press.

———. 1987. *A Thousand Plateaus*. Trans. Brian Massumi. Minneapolis: University of Minnesota Press.

———. 1994. *What Is Philosophy?* Trans. Hugh Tomlinson and Graham Burchell. New York: Columbia University Press.

Detienne, Marcel. 1968. "La Phalange: problèmes et controverses." In *Problèmes de la guerre en Grèce ancienne*, ed. Jean-Pierre Vernant, 119–42. Paris: Ecole des hautes études en science sociale.

de Waal, Frans. 1997. *Good Natured*. Cambridge, Mass.: Harvard University Press.

———. 2006. *Primates and Philosophers: How Morality Evolved*. Princeton, N.J.: Princeton University Press.

Diamond, Jared. 1992. *The Third Chimpanzee: The Evolution and Future of the Human Animal*. New York: HarperCollins.

Di Paolo, Ezequiel. 2005. "Autopoiesis, Adaptivity, Teleology, Agency." *Phenomenology and the Cognitive Sciences* 4, no. 4: 429–52.

Di Paolo, Ezequiel, and Hanne De Jaegher. 2012. "The Interactive Brain Hypothesis." *Frontiers in Human Neuroscience* 6:163. doi:10.3389/fnhum .2012.00163.

Dissanayake, Ellen. 2000. "Antecedents of the Temporal Arts in Early Mother–Infant Interactions." In *The Origins of Music,* ed. N. Wallin, B. Merker, and S. Brown, 389–407. Cambridge, Mass.: MIT Press.

Doyle, Richard. 1997. *On Beyond Living: Rhetorical Transformations in the Life Sciences.* Stanford, Calif.: Stanford University Press.

Drews, Robert. 1993. *The End of the Bronze Age: Changes in Warfare and the Catastrophe ca. 1200 B.C.* Princeton, N.J.: Princeton University Press.

Dreyfus, Hubert. 1972. *What Computers Can't Do.* Cambridge, Mass.: MIT Press.

———. 1992. *What Computers Still Can't Do.* Cambridge, Mass.: MIT Press.

———. 2007. "Why Heideggerian AI Failed, and How Fixing It Would Require Making It More Heideggerian." *Philosophical Psychology* 20, no. 2: 247–68.

Dupuy, Jean-Pierre, and Francisco J. Varela. 1992. "Understanding Origins: An Introduction." In *Understanding Origins,* ed. Francisco Varela and Jean-Pierre Dupuy, 1–26. Dordrecht, Netherlands: Kluwer.

Dyer, J. E. 2011. "The Human Microphone Tactic: Scary or Just Moronic?" *The Optimistic Conservative* (blog), October 9. http://theoptimisticconservative.wordpress.com/2011/10/09.

Edelman, Gerald, and Giulio Tononi. 2000. *A Universe of Consciousness: How Matter Becomes Imagination.* New York: Basic Books.

Ehrenreich, Barbara. 2007. *Dancing in the Streets: A History of Collective Joy.* New York: Metropolitan Books.

Fabing, H. D. 1956. "On Going Berserk: A Neurochemical Inquiry." *The Scientific Monthly* 83, no. 5: 232–37.

Ferguson, Brian. 1995. *Yanomami Warfare: A Political History.* Santa Fe, N.M.: School for American Research Press.

———. 2008. "Ten Points on War." *Social Analysis* 52, no. 2: 32–49.

Ferrill, Arther. 1997. *The Origins of War: From the Stone Age to Alexander the Great.* Boulder, Colo.: Westview Press.

Fine, Cordelia. 2010. *Delusions of Gender: How Our Minds, Society, and Neurosexism Create Difference.* New York: W. W. Norton.

Fletcher, J. 1999. "Using Networked Simulation to Assess Problem Solving by Tactical Teams." *Computers in Human Behavior* 15: 375–402.

Foucault, Michel. 1977. *Discipline and Punish: The Birth of the Prison.* Trans. Alan Sheridan. New York: Vintage Books.

Fracchia, Joseph, and Richard Lewontin. 1999. "Does Culture Evolve?" *History and Theory* 38, no. 4: 52–78.

———. 2005. "The Price of Metaphor." *History and Theory* 44, no. 1: 14–29.

Freeman, Walter J. 2000a. "Emotion Is Essential to All Intentional Behaviors." In *Emotion, Development, and Self-Organization: Dynamic Systems Approaches to Emotional Development,* ed. Marc Lewis and Isabela Granic, 209–35. New York: Cambridge University Press.

———. 2000b. *How Brains Make Up Their Minds.* New York: Columbia University Press.

Fry, Douglas. 2007. *Beyond War: The Human Potential for Peace.* New York: Oxford University Press.

———. 2012. "Life without War." *Science* 336 (May 18): 879–84.

Gabrielsen, Vincent. 2001. "The Social and Economic Impact of Naval Warfare on the Greek Cities." In *War as a Cultural and Social Force: Studies in Ancient Warfare,* ed. T. Bekker-Nielsen and L. Hannestad, 72–89. Copenhagen: Royal Danish Academy of Sciences and Letters.

Gaddis, John Lewis. 2004. *The Landscape of History: How Historians Map the Past.* Oxford: Oxford University Press.

Gallagher, Shaun. 2005. *How the Body Shapes the Mind.* New York: Oxford University Press.

———. 2011. "The Socially Extended Mind." http://www.hum.au.dk/semiotics/docs2/news_archive/2011/shaun-gallagher-masterclasses/the-socially-extended-mind.pdf.

Gallagher, Shaun, and Anthony Crisafi. 2009. "Mental Institutions." *Topoi* 28, no. 1: 45–51.

Gallese, Vittorio. 2001. "The 'Shared Manifold' Hypothesis: From Mirror Neurons to Empathy." *Journal of Consciousness Studies* 8, nos. 5–7: 33–50.

Gallese, Vittorio, C. Keysers, and G. Rizzolatti. 2004. "A Unifying View of the Basis of Social Cognition." *Trends in Cognitive Sciences* 8, no. 9: 396–403.

Gilbert, Scott. 2001. "Ecological Developmental Biology: Developmental Biology Meets the Real World." *Developmental Biology* 233:1–12.

Gomme, A. W. 1933. "A Forgotten Factor of Greek Naval Strategy." *Journal of the Hellenic Society* 53:16–24.

Gray, J. A. 1977. "Drug Effects on Fear and Frustration: Possible Limbic Site of Action of Minor Tranquilizers." In *Handbook of Psychopharmacology.* Vol. 8, *Drugs, Transmitters, and Behavior,* ed. Leslie Iversen, Susan Iversen, and Solomon Snyder, 433–529. New York: Plenum.

Greene, Joshua, and Jonathan Haidt. 2002. "How (and Where) Does Moral Judgment Work?" *Trends in Cognitive Sciences* 6, no. 12: 517–23.

Gregg, Melissa, and Gregory J. Seigworth, eds. 2010. *The Affect Theory Reader.* Durham, N.C.: Duke University Press.

Griffiths, Paul. 1997. *What Emotions Really Are: The Problem of Psychological Categories.* Chicago: University of Chicago Press.

———. 2007. "Evo-Devo Meets the Mind: Toward a Developmental Evolutionary Psychology." In *Integrating Evolution and Development: From Theory to Practice,* ed. Robert Brandon and Roger Sansom, 195–226. Cambridge, Mass.: MIT Press.

Griffiths, Paul, and Russell Gray. 1997. "Replicator II—Judgement Day." *Biology and Philosophy* 12:471–92.

———. 2001. "Darwinism and Developmental Systems." In *Cycles of Contingency: Developmental Systems and Evolution,* ed. Susan Oyama, Paul Griffiths, and Russell Gray, 195–218. Cambridge, Mass.: MIT Press.

———. 2004. "The Developmental Systems Perspective: Organism–Environment Systems as Units of Development and Evolution." In *Phenotypic Integration: Studying the Ecology and Evolution of Complex Phenotypes,* ed. Massimo Pigliucci and Katherine Preston, 409–30. New York: Oxford University Press.

———. 2005. "Discussion: Three Ways to Misunderstand Developmental Systems Theory." *Biology and Philosophy* 20:417–25.

Grossman, David. 1996. *On Killing.* Boston: Little, Brown.

Grosz, Elizabeth. 2008. *Chaos, Territory, Art: Deleuze and the Framing of the Earth.* New York: Columbia University Press.

Hagen, Edward, and Gregory Bryant. 2003. "Music and Dance as a Coalition Signaling System." *Human Nature* 14, no. 1: 21–51.

Hamilton, Edith, and Huntingdon Cairns, eds. 1961. *The Collected Dialogues of Plato.* Princeton, N.J.: Princeton University Press.

Hansen, Mark. 2006. *Bodies in Code: Interfaces with Digital Media.* New York: Routledge.

Hanson, Victor Davis. 1989. *The Western Way of War.* Berkeley: University of California Press.

Hardt, Michael, and Antonio Negri. 2000. *Empire.* Cambridge, Mass.: Harvard University Press.

Harris, William. 2001. *Restraining Rage: The Ideology of Anger Control in Classical Antiquity.* Cambridge, Mass.: Harvard University Press.

Hayles, N. Katherine. 1999. *How We Became Posthuman.* Chicago: University of Chicago Press.

Heidegger, Martin. 1996. *Being and Time.* Trans. Joan Stambaugh. Albany: State University of New York Press.

Hendriks-Jansen, Horst. 1996. *Catching Ourselves in the Act: Situated Activity, Interactive Emergence, Evolution, and Human Thought.* Cambridge, Mass.: MIT Press.

Hoge, C., C. Castro, S. Messer, D. McGurk, D. Cotting, and R. Koffman. 2004. "Combat Duty in Iraq and Afghanistan, Mental Health Problems, and Barriers to Care." *New England Journal of Medicine* 351:13–22.

Hrdy, Sarah. 2009. *Mothers and Others: The Evolutionary Origins of Mutual*

Understanding. Cambridge, Mass.: Harvard University Press.

Hughes, Joe. 2009. *Deleuze's Difference and Repetition: A Reader's Guide.* Edinburgh, U.K.: Edinburgh University Press.

Hurley, Susan. 1998. *Consciousness in Action.* Cambridge, Mass.: Harvard University Press.

———. 2004. "Imitation, Media Violence, and Free Speech." *Philosophical Studies* 117, nos. 1–2: 165–218.

Hutchins, Edwin. 1995. *Cognition in the Wild.* Cambridge, Mass.: MIT Press.

Jablonka, Eva, and Marion J. Lamb. 2005. *Evolution in Four Dimensions: Genetic, Epigenetic, Behavioral, and Symbolic Variation in the History of Life.* Cambridge, Mass.: MIT Press.

Jacobs, Jane. 1970. *The Economy of Cities.* New York: Vintage Books.

Johansson, G. 1976. "Spatio-temporal Differentiation and Integration in Visual Motion Perception: An Experimental and Theoretical Analysis of Calculus-Like Functions in Visual Data Processing." *Psychological Research* 38:379–93.

Jonas, Hans. 2003. *The Phenomenon of Life: Toward a Philosophical Biology.* Evanston, Ill.: Northwestern University Press.

Jones, Graham. 2009. "Solomon Maimon." In *Deleuze's Philosophical Lineage,* ed. Graham Jones and Jon Roffe, 104–29. Edinburgh, U.K.: Edinburgh University Press.

Jordan-Young, Rebecca. 2010. *Brain Storm: The Flaws in the Science of Sex Differences.* Cambridge, Mass.: Harvard University Press.

Joyce, Richard. 2006. *The Evolution of Morality.* Cambridge, Mass.: MIT Press.

Juarrero, Alicia. 1999. *Dynamics in Action: Intentional Behavior as a Complex System.* Cambridge, Mass.: MIT Press.

Kalin, Michael. 2006. "Hidden Pharaohs: Egypt, Engineers, and the Modern Hydraulic." MA thesis, Wolfson College, University of Oxford. http://users.ox.ac.uk/~metheses/Kalin%20Thesis.pdf.

Keeley, Lawrence. 1997. *War before Civilization: The Myth of the Peaceful Savage.* New York: Oxford University Press.

Keller, Evelyn Fox. 2000. *The Century of the Gene.* Cambridge, Mass.: Harvard University Press.

———. 2010. *The Mirage of a Space between Nature and Nurture.* Durham, N.C.: Duke University Press.

Kelly, Raymond. 2000. *Warless Societies and the Origin of War.* Ann Arbor: University of Michigan Press.

Kelso, J. A. Scott. 1995. *Dynamic Patterns: The Self-Organization of Brain and Behavior.* Cambridge, Mass.: MIT Press.

Kilner, Peter. 2000. "Military Leaders' Obligation to Justify Killing in War." Presentation to the Joint Services Conference on Professional Ethics,

Washington, D.C., January 27–28. http://www.au.af.mil/au/awc/awcgate/jscope/kilner00.htm.

Kirkland, Faris. 1995. "Postcombat Reentry." In *War Psychiatry: Textbook of Military Medicine, Part I,* ed. Russ Zajtchuk, 291–317. Washington, D.C.: Office of the Surgeon General.

Lampert, Jay. 2006. *Deleuze and Guattari's Philosophy of History.* London: Continuum.

Lansing, Stephen. 2006. *Perfect Order: Recognizing Complexity in Bali.* Princeton, N.J.: Princeton University Press.

Lawler, Andrew. 2012a. "The Battle over Violence." *Science* 336 (May 18): 829–30.

———. 2012b. "Civilization's Double-Edged Sword." *Science* 336 (May 18): 832–33.

Layton, Julie. 2011. "What Is Waterboarding?" http://science.howstuffworks.com/water-boarding.htm.

Lazenby, J. F. 1985. *The Spartan Army.* Chicago: Bolchazy-Carducci.

LeDoux, Joseph. 1996. *The Emotional Brain.* New York: Simon and Schuster.

Lefebvre, Alexandre. 2008. *The Image of Law: Deleuze, Bergson, Spinoza.* Stanford, Calif.: Stanford University Press.

Lewontin, Richard. 2002. *The Triple Helix: Gene, Organism, and Environment.* Cambridge, Mass.: Harvard University Press.

———. 2005. "The Wars over Evolution." *New York Review of Books,* October 20, 51–54.

Lewontin, Richard, and Richard Levins. 2007. *Biology under the Influence: Dialectical Essays on Ecology, Agriculture, and Health.* New York: Monthly Review Press.

Lifton, Robert. 1973. *Home from the War.* New York: Simon and Schuster.

Lord, Daniel Small. 2007. *On Deep History and the Brain.* Berkeley: University of California Press.

Macedonia, M. 2002. "Games, Simulation, and the Military Education Dilemma." http://www.educause.edu/ir/library/pdf/ffpiu018.pdf.

Mader, Sylvia. 2009. *Biology.* 10th ed. New York: McGraw-Hill.

Maibom, Heidi. 2007. "The Presence of Others." *Philosophical Studies* 132, no. 2: 161–90.

Maimon, Solomon. 2010. *Essay on Transcendental Philosophy.* Trans. Alistair Welchman, Henry Somers-Hall, Merten Reglitz, and Nick Midgley. London: Continuum.

Mallon, Ron, and Stephen Stich. 2000. "The Odd Couple: The Compatibility of Social Construction and Evolutionary Psychology." *Philosophy of Science* 67 (March): 133–54.

Mareschal, Denis, Mark H. Johnson, Sylvain Sirois, Michael W. Spratling, Michael S. C. Thomas, and Gert Westermann. 2007. *Neuroconstruc-*

tivism: How the Brain Constructs Cognition. Vol. 1. New York: Oxford University Press.

Margulis, Lynn. 1998. *Symbiotic Planet*. New York: Basic Books.

Margulis, Lynn, and Dorion Sagan. 1995. *What Is Life?* New York: Simon and Schuster.

Marks, John. 2006. "Molecular Biology in the Work of Deleuze and Guattari." *Paragraph: A Journal of Modern Critical Theory* 29, no. 2: 81–97.

Marshall, Samuel. 1978. *Men against Fire*. Norman: Oklahoma University Press.

Massumi, Brian. 1992. *A User's Guide to Capitalism and Schizophrenia: Deviations from Deleuze and Guattari*. Cambridge, Mass.: MIT Press.

———. 2002. *Parables for the Virtual: Movement, Affect, Sensation*. Durham, N.C.: Duke University Press.

Maturana, Humberto, and Francisco J. Varela. 1980. *Autopoiesis and Cognition: The Realization of the Living*. Boston: Riedel.

May, Todd. 2005. *Gilles Deleuze: An Introduction*. Cambridge: Cambridge University Press.

McAdams, Robert. 1966. *The Evolution of Urban Society*. New York: Aldine.

McBeath, M. K., D. M. Shaffer, and M. K. Kaiser. 1995. "How Baseball Outfielders Determine Where to Run to Catch Fly Balls." *Science* 268, no. 5210: 569–73.

McCarter, M. 2005. "Lights! Camera! Training!" *Military Training Technology* 10, no. 2.

McMenamin, Mark, and Dianna McMenamin. 1994. *Hypersea*. New York: Columbia University Press.

McNeill, William. 1995. *Keeping Together in Time: Dance and Drill in Human History*. Cambridge, Mass.: Harvard University Press.

Menary, Richard, ed. 2010. *The Extended Mind*. Cambridge, Mass.: MIT Press.

Merleau-Ponty, Maurice. 1962. *Phenomenology of Perception*. Trans. Colin Smith. London: Routledge.

Milne, J. G. 1945. "The Economic Policy of Solon." *Hesperia: The Journal of the American School of Classical Studies at Athens* 14, no. 3: 230–45.

Molina, Luis. 1998. "Solon and the Evolution of the Athenian Agrarian Economy." *Pomoerium* 3. http://pomoerium.eu/pomoer/pomoer3/molina.pdf.

Newland, David. n.d. "Vibration of the London Millennium Footbridge." http://www2.eng.cam.ac.uk/~den/ICSV9_06.htm.

Niehoff, Debra. 1999. *The Biology of Violence*. New York: Free Press.

Nisbett, Richard E., and Dov Cohen. 1996. *Culture of Honor: The Psychology of Violence in the South*. Boulder, Colo.: Westview Press.

Noë, Alva. 2004. *Action in Perception*. Cambridge, Mass.: MIT Press.

———. 2009. *Out of Our Heads: Why You Are Not Your Brain, and Other Lessons from the Biology of Consciousness.* New York: Hill and Wang.

Ostrom, Elinor. 2005. "Policies That Crowd Out Reciprocity and Collective Action." In *Moral Sentiments and Material Interests: The Foundations of Cooperation in Economic Life,* ed. Herbert Gintis, Samuel Bowles, Robert Boyd, and Ernst Fehr, 253–75. Cambridge, Mass.: MIT Press.

Otterbein, Keith. 2004. *How War Began.* College Station: Texas A&M Press.

Oyama, Susan. 2000. *The Ontogeny of Information: Developmental Systems and Evolution.* 2nd ed. Durham, N.C.: Duke University Press.

———. 2009. "Friends, Neighbors, and Boundaries." *Ecological Psychology* 21:147–54.

Oyama, Susan, Paul Griffiths, and Russell Gray, eds. 2001. *Cycles of Contingency: Developmental Systems and Biology.* Cambridge, Mass.: MIT Press.

Panksepp, Jaak. 1998. *Affective Neuroscience.* New York: Oxford University Press.

———. 2003. "Damasio's Error?" *Consciousness and Emotion* 4, no. 1: 111–34.

Parkinson, Brian, Agneta Fischer, and Antony Manstead. 2005. *Emotions in Social Relations: Cultural, Group, and Interpersonal Processes.* New York: Psychology Press.

Peterson, Charles H., Sean S. Anderson, Gary N. Cherr, Richard F. Ambrose, Shelly Anghera, Steven Bay, Michael Blum, et al. 2012. "A Tale of Two Spills: Novel Science and Policy Implications of an Emerging New Oil Spill Model." *BioScience* 62, no. 5: 461–69.

Peterson, Dale, and Richard Wrangham. 1997. *Demonic Males: Apes and the Origins of Human Violence.* New York: Houghton Mifflin.

Pierson, D. 1999. "Natural Killers—Turning the Tide of Battle." *Military Review,* May–June, 60–65.

Pigliucci, Massimo. 2010. "Phenotypic Plasticity." In *Evolution: The Extended Synthesis,* ed. Massimo Pigliucci and Gerd B. Müller, 355–78. Cambridge, Mass.: MIT Press.

Pigliucci, Massimo, and Gerd B. Müller, eds. 2010. *Evolution: The Extended Synthesis.* Cambridge, Mass.: MIT Press.

Pringle, Heather. 1998. "The Slow Birth of Agriculture." *Science* 282, no. 5393: 1446–50.

Prinz, Jesse. 2004. *Gut Reactions: A Perceptual Theory of Emotion.* New York: Oxford University Press.

———. 2007. *The Emotional Construction of Morals.* Oxford: Oxford University Press.

———. 2012. *Beyond Human Nature: How Culture and Experience Shape Our Lives.* London: Allen Lane.

Pritchett, W. Kendrick. 1974. *The Greek State at War.* Part II. Berkeley: University of California Press.

Protevi, John. 1990. "The *Sinnsfrage* and the *Seinsfrage*." *Philosophy Today* 34, no. 4: 321–33.

———. 1998. "The 'Sense' of 'Sight': Heidegger and Merleau-Ponty on the Meaning of Bodily and Existential Sight." *Research in Phenomenology* 28:211–23.

———. 2001. *Political Physics: Deleuze, Derrida, and the Body Politic.* London: Athlone/Continuum.

———. 2006. "Deleuze, Guattari, and Emergence." *Paragraph: A Journal of Modern Critical Theory* 29, no. 2: 19–39.

———. 2009. *Political Affect: Connecting the Social and the Somatic.* Minneapolis: University of Minnesota Press.

———. 2010. "Deleuze and Wexler: Thinking Brain, Body, and Affect in Social Context." In *Cognitive Architecture: From Biopolitics to Noopolitics—Architecture and Mind in the Age of Communication and Information,* ed. Deborah Hauptmann and Warren Neidich, 168–83. Rotterdam: 010 Publishers.

———. 2011. "Impossible Demands for 'Proof' in the Giffords Assassination Case." *The Contemporary Condition* (blog), January 10. http://contemporarycondition.blogspot.com/2011/01/impossible-demands-for-proof-in.html.

Putnam, Hilary. 1975. *Mind, Language, and Reality.* Vol. 2 of *Philosophical Papers.* Cambridge: Cambridge University Press.

Raaflaub, Kurt, Josiah Ober, and Robert Wallace. 2007. *Origins of Democracy in Ancient Greece.* Berkeley: University of California Press.

Read, Jason. 2010. "The Production of Subjectivity: From Transindividuality to the Commons." *New Formations: A Journal of Culture/Theory/Politics* 70:113–31.

Reiser, Marc. 1993. *Cadillac Desert: The American West and Its Disappearing Water.* New York: Penguin.

Richerson, Peter, and Robert Boyd. 2005. *Not by Genes Alone: How Culture Transformed Human Evolution.* Chicago: University of Chicago Press.

Ristic, Igor. 2011. "The Human Microphone #OccupiesWallStreet." *CharacteRistic* (blog), October 11. http://igorristic.wordpress.com/2011/10/11/the-human-microphone-occupieswallstreet/.

Rodriguez-Trelles, Francisco, Rosa Tarrio, and Francisco Ayala. 2005. "Is Ectopic Expression Caused by Deregulatory Mutations or due to Gene-Regulation Leaks with Evolutionary Potential?" *BioEssays* 27:592–601.

Rosch, Eleanor. 1978. "Principles of Categorization." In *Cognition and Categorization,* ed. Eleanor Rosch and Barbara Lloyd, 27–48. Hillsdale, N.J.: Lawrence Erlbaum Associates.

Rosen, Robert. 1991. *Life Itself: A Comprehensive Inquiry into the Nature, Origin, and Fabrication of Life.* New York: Columbia University Press.

Rudrauf, David, Antoine Lutz, Diego Cosmelli, Jean-Philippe Lachaux, and Michel Le Van Quyen. 2003. "From Autopoiesis to Neurophenom-enology: Francisco Varela's Exploration of the Biophysics of Being." *Biological Research* 36, no. 1: 27–65.

Runciman, W. G. 1998. "Greek Hoplites, Warrior Culture, and Indirect Bias." *Journal of the Royal Anthropological Institute* 4, no. 4: 731–51.

———. 2005a. "Culture Does Evolve." *History and Theory* 44, no. 1: 1–13.

———. 2005b. "Rejoinder to Fracchia and Lewontin." *History and Theory* 44, no. 1: 30–41.

Sacks, David. 1995. *A Dictionary of the Ancient Greek World.* New York: Oxford University Press.

Satz, Debra, and John Ferejohn. 1994. "Rational Choice and Social Theory." *The Journal of Philosophy* 91, no. 2: 71–87.

Sauvagnargues, Anne. 2004. *De l'animal à l'art.* In *La Philosophie de Deleuze,* 117–227. Paris: Presses Universitaires de France.

———. 2005. *Deleuze et l'art.* Paris: Presses Universitaires de France.

———. 2009. *Deleuze: L'empirisme transcendental.* Paris: Presses Universitaires de France.

Schwartz, Barry, Richard Schuldenfrei, and Hugh Lacey. 1979. "Operant Psychology as Factory Psychology." *Behaviorism* 6:229–54.

Sharp, Hasana. 2011. *Spinoza and the Politics of Renaturalization.* Chicago: University of Chicago Press.

Shaviro, Steven. 2009. *Without Criteria: Kant, Whitehead, Deleuze, and Aesthetics.* Cambridge, Mass.: MIT Press.

Shay, Jonathan. 1994. *Achilles in Vietnam.* New York: Macmillan.

Sides, John. 2011. "Political Vitriol and Political Violence." *The Monkey Cage* (blog), January 9. http://themonkeycage.org/blog/2011/01/09/political_vitriol_and_politica/.

Simondon, Gilbert. 1995. *L'Individu et sa genèse physico-biologique.* Grenoble, France: Jérôme Millon.

Singer, T., B. Seymour, J. O'Doherty, H. Kaube, R. J. Dolan, and C. Frith. 2004. "Empathy for Pain Involves the Affective but Not Sensory Components of Pain." *Science* 303 (February 20): 1157–62.

Skrbina, David. 2005. *Panpsychism in the West.* Cambridge, Mass.: MIT Press.

———, ed. 2009. *Mind That Abides: Panpsychism in the New Millennium.* Amsterdam: John Benjamins.

Smith, Daniel W. 2009. "Genesis and Difference: Deleuze, Maimon, and the Post-Kantian Reading of Leibniz." In *Deleuze and the Fold: A Critical Reader,* ed. Sjoerd van Tuinen and Niamh McDonnell, 132–54. Basingstoke, U.K.: Palgrave Macmillan.

Sober, Elliott, and David Sloan Wilson. 1998. *Unto Others: The Evolution and Psychology of Unselfish Behavior.* Cambridge, Mass.: Harvard University Press.

Speidel, Michael. 2002. "Berserks: A History of Indo-European 'Mad Warriors.'" *Journal of World History* 13, no. 2: 253–90.

Sponsel, Leslie. 2000. "Response to Otterbein." *American Anthropologist* 102, no. 4: 837–41.

Ste. Croix, G. E. M. de. 1972. *The Origins of the Peloponnesian War.* Ithaca, N.Y.: Cornell University Press.

———. 1981. *The Class Struggle in the Ancient Greek World.* Ithaca, N.Y.: Cornell University Press.

———. 2004. *Athenian Democratic Origins and Other Essays.* New York: Oxford University Press.

Steinbock, Anthony J. 1995. *Home and Beyond: Generative Phenomenology after Husserl.* Evanston, Ill.: Northwestern University Press.

Sterelny, Kim, and Paul Griffiths. 1999. *Sex and Death: An Introduction to the Philosophy of Biology.* Chicago: University of Chicago Press.

Stern, Daniel. 1985. *The Interpersonal World of the Infant.* New York: Basic Books.

Strawson, Galen. 2006. "Realistic Monism: Why Physicalism Entails Panpsychism." *Journal of Consciousness Studies* 13, nos. 10–11: 3–31.

Thompson, Evan. 2001. "Empathy and Consciousness." *Journal of Consciousness Studies* 8, nos. 5–7: 1–32.

———. 2007. *Mind in Life: Biology, Phenomenology, and the Sciences of Mind.* Cambridge, Mass.: Harvard University Press.

Thompson, Evan, and Francisco J. Varela. 2001. "Radical Embodiment: Neural Dynamics and Consciousness." *Trends in Cognitive Sciences* 5, no. 10: 418–25.

Toch, Hans. 1992. *Violent Men: An Inquiry into the Psychology of Violence.* Washington, D.C.: American Psychological Association.

Toscano, Alberto. 2006. *The Theatre of Production: Philosophy and Individuation between Kant and Deleuze.* London: Palgrave Macmillan.

Trevarthen, Colin. 1999. "Musicality and the Intrinsic Motive Pulse: Evidence from Human Psychobiology and Infant Communication." *Musicae Scientiae* (Special Issue): 155–215.

Turetsky, Phil. 2004. "Rhythm, Assemblage, and Event." In *Deleuze and Music,* ed. Ian Buchanan and Marcel Swoboda, 140–58. Edinburgh, U.K.: Edinburgh University Press.

Turner, J. Scott. 2000. *The Extended Organism: The Physiology of Animal-Built Structures.* Cambridge, Mass.: Harvard University Press.

Ulmen, G. L. 1978. *The Science of Society: Toward an Understanding of the Life and Work of Karl August Wittfogel.* The Hague: Mouton.

U.S. Department of Homeland Security. 2009. "Rightwing Extremism: Current Economic and Political Climate Fueling Resurgence in Radicalization and Recruitment." http://www.fas.org/irp/eprint/rightwing.pdf.

van der Kolk, Bessel, and M. Greenberg. 1987. "The Psychobiology of the

Trauma Response." In *Psychological Trauma*, ed. Bessel van der Kolk, 63–87. Washington, D.C.: American Psychiatric Press.

Varela, Francisco J. 1976. "Not One, Not Two." *The CoEvolution Quarterly*, Fall, 62–67.

———. 1977. "On Being Autonomous: The Lessons of Natural History for Systems Theory." In *Applied Systems Research*, ed. George J. Klir, 77–85. New York: Plenum Press.

———. 1979a. *Principles of Biological Autonomy*. New York: North Holland.

———. 1979b. "Reflections on the Chilean Civil War." *Lindisfarne Letter* 8 (Winter): 13–19.

———. 1981. "Describing the Logic of the Living: The Adequacy and Limitations of the Idea of Autopoiesis." In *Autopoiesis: A Theory of Living Organization*, ed. Milan Zeleny, 36–48. New York: North Holland.

———. 1991. "Organism: A Meshwork of Selfless Selves." In *Organism and the Origins of Self*, ed. Alfred I. Tauber, 79–107. The Hague: Kluwer.

———. 1992a. "Making It Concrete: Before, During, and After Breakdowns." In *Revisioning Philosophy*, ed. James Ogilvy, 97–111. Albany: State University of New York Press.

———. 1992b. "Whence Perceptual Meaning? A Cartography of Current Ideas." In *Understanding Origins*, ed. Francisco Varela and Jean-Pierre Dupuy, 235–63. Dordrecht, Netherlands: Kluwer.

———. 1995. "Resonant Cell Assemblies: A New Approach to Cognitive Functions and Neuronal Synchrony." *Biological Research* 28:81–95.

———. 1996. "Neurophenomenology: A Methodological Remedy for the Hard Problem." *Journal of Consciousness Studies* 3, no. 4: 330–49.

———. 1999a. *Ethical Know-How: Action, Wisdom, and Cognition*. Stanford, Calif.: Stanford University Press.

———. 1999b. "The Specious Present: A Neurophenomenology of Time Consciousness." In *Naturalizing Phenomenology: Issues in Contemporary Phenomenology and Cognitive Science*, ed. Jean Petitot, Francisco J. Varela, Bernard Pachoud, and Jean-Michel Roy, 266–314. Stanford, Calif.: Stanford University Press.

———. 1999c. "Steps to a Science of Inter-being: Unfolding the Dharma Implicit in Modern Cognitive Science." In *The Psychology of Awakening: Buddhism, Science, and Our Day-to-Day Lives*, ed. Gay Watson, Stephen Batchelor, and Guy Claxton, 71–89. New York: Random House.

———. 2002. "Autopoïese et émergence." In *La Complexité, vertiges et promesses*, ed. Réda Benkirane, 159–76. Paris: Le Pommier.

Varela, Francisco J., and Antonio Coutinho. 1991. "Immunoknowledge: The Immune System as a Learning Process of Somatic Individuation." In *Doing Science: The Reality Club*, ed. John Brockman, 237–56. New York: Prentice Hall.

Varela, Francisco J., and Natalie Depraz. 2005. "At the Source of Time: Valence and the Constitutional Dynamics of Affect." *Journal of Consciousness Studies* 12, nos. 8–10: 61–81.

Varela, Francisco J., Humberto Maturana, and R. Uribe. 1974. "Autopoiesis: The Organization of the Living, Its Characterization and a Model." *BioSystems* 5, no. 4: 187–96.

Varela, Francisco J., Evan Thompson, and Elizabeth Rosch. 1991. *The Embodied Mind.* Cambridge, Mass.: MIT Press.

Varela, Francisco J., F. J. Lachaux, J.-P. Rodriguez, and J. Martinerie. 2001. "The Brainweb: Phase Synchronization and Large-Scale Integration." *Nature Reviews: Neuroscience* 2:229–39.

Villani, Arnaud. 1999. *La guêpe et l'orchidée: Essai sur Gilles Deleuze.* Paris: Belin.

Vogel, Lise. 2000. "Domestic Labor Revisited." *Science and Society* 64, no. 2: 151–70.

Ward, Lee. 2001. "Nobility and Necessity: The Problem of Courage in Aristotle's Nicomachean Ethics." *The American Political Science Review* 95, no. 1: 71–83.

Watt, Douglas. 2000. "Emotion and Consciousness: Part II: A Review of Antonio Damasio's *The Feeling of What Happens.*" *Journal of Consciousness Studies* 7, no. 3: 72–84.

Weizman, Eyal. 2007. *Hollow Land: Israel's Architecture of Occupation.* London: Verso.

Welchman, Alistair. 2009. "Deleuze's Post-critical Metaphysics." *Symposium* 13, no. 2: 25–54.

Welton, Donn. 2000. *The Other Husserl: The Horizons of Transcendental Phenomenology.* Bloomington: Indiana University Press.

Welton, Donn, ed. 2003. *The New Husserl: A Critical Reader.* Bloomington: Indiana University Press.

West-Eberhard, Mary Jane. 2003. *Developmental Plasticity and Evolution.* New York: Oxford University Press.

———. 2007. "Dancing with DNA and Flirting with the Ghost of Lamarck." *Biology and Philosophy* 22:439–51.

Wetherell, Margaret. 2012. *Affect and Emotion: A New Social Science Understanding.* London: Sage.

Wexler, Bruce. 2006. *Brain and Culture: Neurobiology, Ideology, and Social Change.* Cambridge, Mass.: MIT Press.

Wheeler, Michael. 1997. "Cognition's Coming Home: The Reunion of Mind and Life." In *Proceedings of the Fourth European Conference on Artificial Life,* ed. Phil Husbands and Inman Harvey, 10–19. Cambridge, Mass.: MIT Press.

———. 2005. *Reconstructing the Cognitive World.* Cambridge, Mass.: MIT Press.

——. 2007. "Traits, Genes, and Coding." In *Philosophy of Biology*, ed. Mohan Matthen and Christopher Stephens, 369–99. Amsterdam: North Holland.

——. 2010. "In Defense of Extended Functionalism." In *The Extended Mind*, ed. Richard Menary, 245–70. Cambridge, Mass.: MIT Press.

——. 2011. "Mind in Life or Life in Mind? Making Sense of Deep Continuity." *Journal of Consciousness Studies* 18, nos. 5–6: 148–68.

Williams, James. 2003. *Gilles Deleuze's Difference and Repetition: A Critical Introduction and Guide*. Edinburgh, U.K.: Edinburgh University Press.

——. 2005. *The Transversal Thought of Gilles Deleuze: Encounters and Influences*. Manchester, U.K.: Clinamen Press.

——. 2008. *Gilles Deleuze's Logic of Sense: A Critical Introduction and Guide*. Edinburgh, U.K.: Edinburgh University Press.

Wittfogel, Karl. 1957. *Oriental Despotism: A Comparative Study of Total Power*. New Haven, Conn.: Yale University Press.

Wolpert, Lewis. 1969. "Positional Information and the Spatial Pattern of Cellular Differentiation." *Journal of Theoretical Biology* 25, no. 1: 1–47.

——. 1989. "Positional Information Revisited." *Development* 107 (April): 3–12.

Wood, Denis. 2004. *Five Billion Years of Global Change*. New York: Guilford.

Worster, Donald. 1985. *Rivers of Empire: Water, Aridity, and the Growth of the American West*. New York: Oxford University Press.

Young, Iris Marion. 1990. *Throwing Like a Girl and Other Essays in Feminist Philosophy and Social Theory*. Bloomington: Indiana University Press.

——. 2005. *On Female Body Experience: "Throwing Like a Girl" and Other Essays*. New York: Oxford University Press.

Young, Rebecca L., and Alexander V. Badyaev. 2007. "Evolution of Ontogeny: Linking Epigenetic Modeling and Genetic Adaptation in Skeletal Structures." *Integrative and Comparative Biology* 47, no. 2: 234–44.

Zahavi, Dan. 2005. *Subjectivity and Selfhood: Investigating the First-Person Perspective*. Cambridge, Mass.: MIT Press.

Publication History

An earlier version of "Introduction II" appeared as "Beyond Auto-poiesis: Inflections of Emergence and Politics in Francisco Varela," in *Emergence and Embodiment: New Essays on Second-Order Systems Theory,* ed. Bruce Clarke and Mark Hansen, 94–112 (Durham, N.C.: Duke University Press, 2009). Reprinted by permission of Duke University.

An earlier version of chapter 1 appeared as "Geophilosophy and Hydro-bio-politics," in *Deleuze and History,* ed. Jeffrey Bell and Claire Colebrook, 92–102 (Edinburgh, U.K.: Edinburgh University Press, 2008), http://www.euppublishing.com/.

An earlier version of chapter 2 appeared as "Affect, Agency, and Responsibility: The Act of Killing in Contemporary Warfare," *Phenomenology and the Cognitive Sciences* 7, no. 3 (2008): 405–13.

An earlier version of chapter 3 appeared as "Rhythm and Cadence, Frenzy and March: Music and the Geo-bio-techno-affective Assemblages of Ancient Warfare," *Theory and Event* 13, no. 3 (2010), doi:10.1353/tae.2010.0006. Copyright 2010 by John Protevi and the Johns Hopkins University Press. Reprinted with permission of the Johns Hopkins University Press.

An earlier version of chapter 4 appeared as "Deleuze and Wexler: Thinking Brain, Body, and Affect in Social Context," in *Cognitive Architecture: From Biopolitics to Noopolitics—Architecture and Mind in the Age of Communication and Information,* ed. Deborah Hauptmann and Warren Neidich, 168–83 (Rotterdam, Netherlands: 010 Publishers, 2010).

An earlier version of chapter 5 appeared as "Adding Deleuze to the Mix," *Phenomenology and the Cognitive Sciences* 9, no. 3 (2010): 417–36.

Index

John Protevi is Phyllis M. Taylor Professor of French Studies and professor of philosophy at Louisiana State University. He is the author of *Political Affect: Connecting the Social and the Somatic* (Minnesota, 2009), *Time and Exteriority* (1994), and *Political Physics* (2001); the coauthor of *Deleuze and Geophilosophy* (with Mark Bonta; 2004); and the editor of *A Dictionary of Continental Philosophy* (2006).